D0945426

Understanding Scientific Literatures: A Bibliometric Approach

The MIT Press
Cambridge, Massachusetts, and London, England

Understanding Scientific Literatures: A Bibliometric Approach

Joseph C. Donohue

This book was composed at The MIT Press on the Imlac PDS-1 and compiled by Composition Technology, Inc. for output on the Harris Fototronic CRT at The Colonial Press. It was printed on Pinnacle Book Offset and bound by The Colonial Press Inc. in the United States of America.

Library of Congress Cataloging in Publication Data
Donohue, Joseph C.

Understanding Scientific Literatures

Bibliography: p.
1. Information science. 2. Library science.
I. Title. II. Title: Bibliometric approach.
Z669.8.D65 020 72-10334
ISBN 0-262-04039-5

To my teachers and colleagues at Case Western Reserve—
with respect and affection

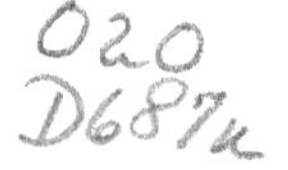

Contents

Preface

As the author states, the methods he describes treat the growth of a scientific literature as a social phenomenon in its own right, not as a material byproduct of the knowledge and concepts of the science written about. Management of the literature of physics, say, does not first demand the study of physics, but of physicists. Management of the literature of information sciences demands first the study of information scientists.

For this we need to be able to recognize information scientists and the literature of information sciences. In terms of subject matter the criteria for recognition are at present in some dispute. In terms just of reading and writing there is no need for dispute. The literature of the information sciences is what is common to what is read and written by information scientists. Information scientists are those who read and write the literature of the information scientists.

This is not absurd, still less is it useless. If such a symbiosis between a group of people and their recorded discourse persists, it will develop characteristic ways of reading, writing, and requesting. We can observe these as purely social, bibliographic, events. Provided that we have observed correctly the correct events and know something of the social patterns of like events, we can detect early symptoms of emerging topics and disciplines. Moreover we can make reasonable managerial decisions to anticipate future growth of demand and of obsolescence and can develop effective retrieval tools and languages.

Over the past quarter century quantitative study of bibliographic behavior in its own right has expanded greatly. For studies of this kind Alan Pritchard suggested in 1969 the name "bibliometrics," from considerations of terminology in parallel fields and etymology in the past of this one. Bibliometric studies have been made for a long time mainly, as is surprising and proper, by librarians: Surprising because librarians are noted neither for spare time nor spare money; proper because the key phenomena of any field are most likely to be suspected and detected by day to day workers in it. Exploitation and explanation of the phenomena are then open to anyone with the understanding, skills, and opportunities to exploit or explain.

The key phenomenon applied by the author is that usually called "Bradford's law" after the British librarian Samuel Clement Bradford (1878–1948) whose insight suspected it and industry displayed it. Those who knew him are glad that his name is used to label it.

Like most social laws, Bradford's law differs from Ohm's law, say, or the Ten Commandments. It is a statistical law, referring to the usual overall behavior of collections, not to the precise behavior of individuals. Nowadays we know that it holds over a very wide range of bibliographical situations but are not clear why it should do so. Understanding of this will unify some hitherto stubbornly empirical data and throw much light on the aims and concepts proper to the information sciences. But fortunately we do not have to wait for complete understanding before we can use it, any more than we need wait for a complete explanation of human psychology before shouting "Fire!"

Here Bradford behavior is shown as the root of techniques for solving a particular problem, the deeper and wider significances being treated in less detail. For those who wish to follow these up, the author gives adequate references, and there can be no better introduction to the theory of a phenomenon than to have applied it in practice.

Because the phenomena are statistical, their discussion cannot avoid being to some extent mathematical. Plain mathematics should need apology no more than should plain English. But at the moment any group of people concerned with the management of messages can display such wide variations of the balance of numeracy to literacy that any presentation of mathematics must be in some sort defective. Always there will be some who take fright at any nonlexical configuration of letters and symbols and some who lapse into generalized Fourier analysis in midsentence. For those in between, probably the most useful approach is to outline the chain of argument in words, illustrate it with mathematical expressions that are simple without being uncouth, and give enough references. These the author has done.

The publication of this book gives me much satisfaction, because it marks the end of the infancy of bibliometrics, during which I have been around largely as a spectator. I knew Bradford in his last years, I know the author, and I know and have spent a happy time, for me anyway, in the same invigorating academic environment. Thus I have not only watched people sow for others to reap but also am now able to do the reaping. May others do so, and do some sowing also.

Robert A. Fairthorne
February 1972

Acknowledgments

The reader who is familiar with William Goffman's contributions to information science theory will readily recognize how basic his thinking is to the study presented here. In the preparation of this work Dr. Goffman has given most generously of his time, ideas, and encouragement. Professor Tefko Saracevic has followed the study with a kindly interest and has made many valuable suggestions.

Many other people have contributed to making this work possible. Financial support during my years at Case Western Reserve University was provided under fellowships from the United States Office of Education and the AHS Foundation. Professor Alan Rees first encouraged me to embark on the doctoral program at Reserve, and during my stay there he contributed greatly to my thinking, as did my other teachers and my colleagues. Dr. Conrad Rawski first directed my attention to some important differences between knowledge of a subject and knowledge of its literature, distinctions that are central to this study.

The hundreds of publications examined were collected with the courteous assistance of many librarians at Case Western Reserve University and the University of Maryland. Especially helpful were the staff members of the Allen Memorial Medical Library where much of the work was done. Copies of some elusive items were provided by the Directorate of Information Science, Air Force Office of Scientific Research (AFOSR).

Miss Carole Armstrong and Mr. Sam Markson very kindly provided me with citation data compiled for a study that they made on the relation of author productivity to citations. Staff members of the Case Western Reserve University Documentation Center were very helpful in the preparation of machine-readable data. I am especially grateful to Mr. Frederick Leise, who did much of the data processing, to Mrs. Lissa Soergel, who provided editorial assistance, and to Mrs. Carole Peppi, who exercised great care and patience in typing many successive drafts of the manuscript. Dr. Howard Roberts read and commented on sections of the final draft, and Miss Helen Reynolds provided a final proofreading.

Finally, I want to thank Betty Donohue, who has been a continuing source of encouragement.

Joseph C. Donohue

1
Introduction

1.1 Subject Literatures: Some Problems in Defining Them

A literature is a body of thought as expressed in published writings. A basic function of the library is to collect the publications that record literatures for future use. At one time in history, the total body of recorded literature was small enough that a library might aspire to collect all of it. The cost and scarcity of books and other printed matter was such that libraries could collect indiscriminately; any published literature could be presumed to have some value. It is no longer economically possible to accept that presumption today. The weary householder emptying the Sunday newspaper into the trash is well aware of the small proportion of *news*paper to waste paper that he receives. The academic or corporate scientist may doubt the value of much of research publication, even his own, but with similar resignation he continues to publish. Once, librarians prided themselves on the sheer size of their collections; even today, college catalogs boast the number of library accessions as though somehow it were better to have an enormous number of useless items than a small number of pertinent ones. The more sophisticated and more honest librarian today looks for some means of tailoring his library's collection to the current and potential needs of its users. There are many reasons for doing so. The useful portions of a collection can be made more readily accessible if the chaff is diminished. Further, the sheer force of economics makes it increasingly difficult to contain within one library all the publications that may conceivably be of value to its users. To eliminate a publication from a library is not necessarily a disparagement of the material in question.

There are, to be sure, libraries "of record" that attempt to collect *all* publications within a given subject scope. Whatever may be the ultimate social value of this heroic effort, it is certain that the effort itself becomes continually more difficult and costly, perhaps prohibitively so. Contrary to a popularly held myth, even the massive Library of Congress does not retain all of the books it receives under the copyright provisions. With smaller libraries, the case is more obvious. All responsibly operated libraries collect according to a conscious plan that delimits the scope of acquisitions. In most large systems, formal acquisitions policies have been written and promulgated at one time or another, but these policies are most often expressed in purely qualitative terms and are administered in a way that leaves a great deal to intuitive, subjective judgment. In some instances, attempts are made to render selection more objective by substituting the intuitive judgments of a committee for those of an individual. While this method may result in a more balanced collection of library materials, it is not a step toward an objective manner of selection. Further, because this subjective method is based on the hindsight of the selectors, it offers no hope of predicting what tomorrow's needs will be.

1.2 A Method for the Identification of Literatures

The study presented here is motivated by the conviction that in order to operate effectively, libraries must identify literatures of high utility to their clienteles and must acquire and organize these literatures in such a way as to optimize their usefulness. It is assumed that effectiveness and economy cannot be developed by relying on informal and intuitive selection procedures.

In recent years, formal analytic and predictive techniques have been developed for the study of subject literatures, as part of a growing, if primitive, discipline called "information science," which attempts an understanding of knowledge and particularly of its diffusion.* Certain of these techniques have been chosen, which have been used previously to elucidate the nature of one or another subject literature. These include

A technique for the identification of a "core" literature—that segment which is potentially most useful,

A method for ranking publications in zones of diminishing importance,

A means of establishing a transition point between zones of higher and lower utility,

Techniques for tracing the spread of ideas, analogous to the statistical techniques used in the study of epidemics; and,

A method for classifying segments of a literature by reference to the interconnections shown in the citations given by publications.

These varied approaches, each of them previously applied to different subject literatures, are brought together here and focused on the literature of a single multidisciplinary field, information science—or more correctly, on a segment of that literature. As a group, these techniques may be referred to as *bibliometric*, following the suggestion of A. Pritchard, who outlines the reasons for preferring this term over the earlier-used expression *statistical bibliography*. Pritchard notes that though introduced as early as 1922 by E. Wyndham Hulme,

*The techniques are formal in the sense that they rely on a study of the formal characteristics of published literatures rather than intellectual content. They are also formal in the sense of being operational.

that phrase never achieved general acceptance. Several definitions are given of statistical bibliography:

> To shed light on the processes of written communication and of the nature and course of development of a discipline (in so far as this is displayed through written communication), by means of counting and analyzing the various facets of written communication.
>
> The assembling and interpretation of statistics relating to books and periodicals . . . to demonstrate historical movements, to determine the national or universal research use of books and journals, and to ascertain in many local situations the general use of books and journals. [Pritchard 1969b]

Pritchard finds these definitions useful, but considers the phrase statistical bibliography unsatisfactory for several reasons:

> The term is clumsy, not very descriptive, and can be confused with statistics itself or bibliographies on statistics. This latter point was made by M. G. Kendall upon receiving a copy of my paper [Pritchard 1969a] and he suggested that the name of the subject be changed. This provided the final impetus for this note.
>
> Therefore it is suggested that a better name for this subject (as previously defined) is *bibliometrics*, i.e., the application of mathematics and statistical methods to books and other media of communication. An intensive search of the literature has failed to reveal any previous use of this term and an approach to the OED again failed to find that the term had been used before.
>
> The beauty of this term is that, whilst this particular combination is a neologism and therefore to be treated with a certain amount of suspicion, it has very close links to the accepted, and analogous 'biometrics', 'econometrics', and 'scientometrics'. The latter term is a Russian one for the application of quantitative methods to the history of science and obviously overlaps with bibliometrics to a considerable extent. *En passant*, it is greatly to be regretted that the very logical Russian term for studies of all types on the processes of science 'scientology' has such unfortunate connotations in the West. [Pritchard 1969b].

In the present study, the accent is placed on the development of an approach that integrates several techniques. No great significance should be attached to the choice of the particular subject literature selected for the test. That choice was partly fortuitous and partly the result of the investigator's continuing interest in heuristic studies, which resulted in the ready availability of many of the publications involved. It would be unwise to make sweeping statements about the nature of the development of information science, based on the study of this sample, for two reasons. First, the sample size is small and covers a short period of time. Second, the primary corpus of publications was deliberately chosen by an arbitrary criterion, namely, that it represented the literature resulting from one sponsor's program in information science. Some such arbitrary limit was essential; it should be clear that no one program represents the entirety of a science of information. The study does tell us some interesting things about information science. But much more importance is placed by the investigator on the problem of bringing together different techniques into a general method for the analysis of a subject literature.

Two assumptions are basic to the present study: (1) that there are regularities observable in patterns of authorship, publication, and citation of literatures, and (2) that careful observation of these regularities can yield principles for the effective management of the literature. It is not enough, however, to observe regularities. The observed phenomena must be *measured* if we are to gain power to predict and/or control. Hence the approach used here, which has been termed *bibliometric*, that is, the "quantitative treatment of the properties of recorded discourse and behaviour appertaining to it." [Fairthorne 1969, p. 319].

The results of the study presented here indicate validity in the method developed and hold promise that with a great deal more work it may be made the basis for an effective and economical acquisitions policy.

1.3 Organization of the Study

This study is arranged according to the following scheme: First, an examination is made of some basic notions regarding (1) human information and communication processes, especially those that pertain to the management of recorded information; (2) the relationships found among certain functional roles, as for example, between scientist and librarian in the production and use of knowledge; and (3) the relationships among information science, library science, and subject disciplines. Second, a review is presented of certain information-scientific techniques that have been applied previously to the formal aspects of subject literatures as distinguished from the conceptual content of these literatures. Third, a general method is developed for using these techniques in the analysis of various aspects of a single literature. Fourth, the method is applied to a defined segment of the literature of information science (IS). Fifth, results are analyzed and directed toward the discovery of basic processes observable in the literature examined, and an explication of the method's potential application to other literatures. Sixth, implications are suggested of the method for the management of library collections, especially as it may be applied to problems of acquisition and organization of documentary materials.

1.4 Literatures and Libraries: Some Concepts Basic to This Study

As used in this study, a *literature* is a body of thought about a given subject as expressed in published writings; a *library* is an agency for the management of human records, especially the published writings of the respective literatures. It is this management function, directed toward facilitating the eventual use of the records that makes the library more than a mere storage place and makes the librarian something other than merely a conservator of physical artifacts. *Librarianship* is one of several occupations that are directed to the management of records. (See Figure 1.1.) Its theoretical basis has been sought under the rubric of *library science*, a quasi-discipline that might better be called "library studies" because it is notably weak in

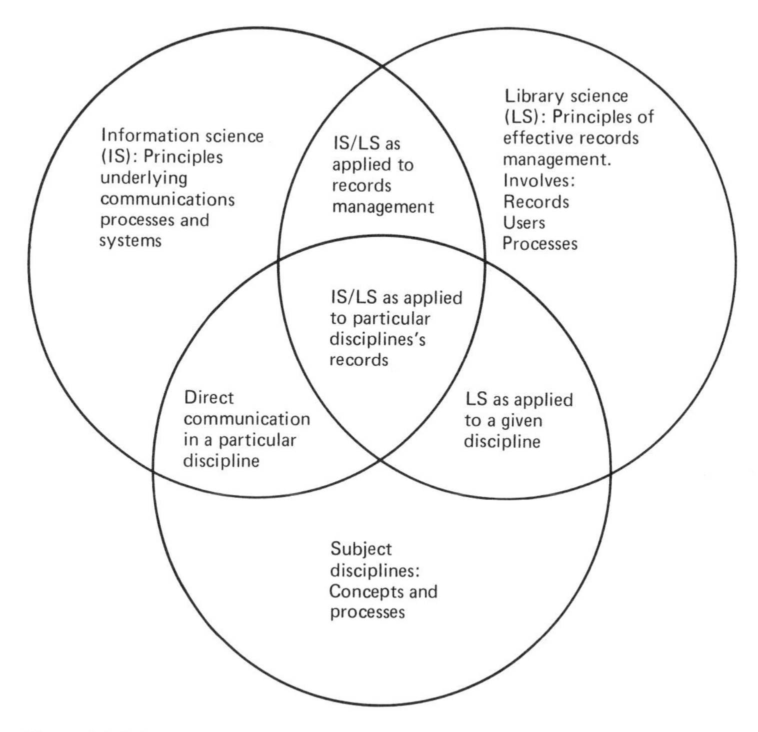

Figure 1.1 Information science, library science, and subject disciplines.

scientific rigor. In recent years library science and other disciplines have interacted frequently under the name *information science*, which has been defined as "the study of the principles underlying information processes and systems" (Goffman 1970). Such study is not of itself new; it has antecedents reaching back to the classical philosophers' concern with epistemology. What is new is the attempt to create a coherent, formal discipline that has as its central concern the elucidation of these underlying principles and their application to problems in a wide range of endeavors.

In the present work several techniques drawn from information science are identified as being especially pertinent to the problems of libraries, and these techniques are applied to the analysis of a single literature. The intent is not merely to illustrate a range of analytical techniques now available to libraries in managing collections but rather to begin a synthesis of them, showing relationships among the elements of new knowledge that they elicit concerning the structure of the literature and the processes occurring within it. It is this understanding of literature forms and processes, rather than the understanding of the literatures' ideational content that seems to offer the most useful means for prediction and control in the management of library resources. This is not at all to deny that a literature itself is ultimately of value for the ideas it contains. But the library does not deal directly with ideas; it makes its contribution to the world of ideas through its effective management of the *records* of ideas.*

*The same may be said for archives and museums. The differences among libraries, archives, museums, and managerial information systems, whether computerized or not, though important, are less important than the similarities in basic purpose and functions.

Librarians today emphasize their role as *communicators* rather than as *records managers*, even to the extent that they sometimes appear to favor neglecting the record itself. This shift in emphasis may have both positive and negative results. On the positive side, it highlights the librarians' concern for the whole communication process rather than solely with the handling of artifacts, and this may increase their awareness and society's awareness of the social and psychological processes through which libraries perform their service. But the broadening of his role could also vitiate the librarian's potential contribution if, by taking as his domain too many aspects of communication, he fails to define for himself an area of competence. The present work is prompted by the conviction that the most important contribution of the librarian as such lies in the competent management of human records. The role of records manager in this sense is not the shallow clerical one that the term may conjure. It implies a deep understanding of the record, of the user's need for the record, and of the communication processes by which the user makes effective contact with it (see Figure 1.2).

Some students of information science hold that the librarian as records manager must have the same kind of understanding of the records' contents as has the user. Others distinguish between

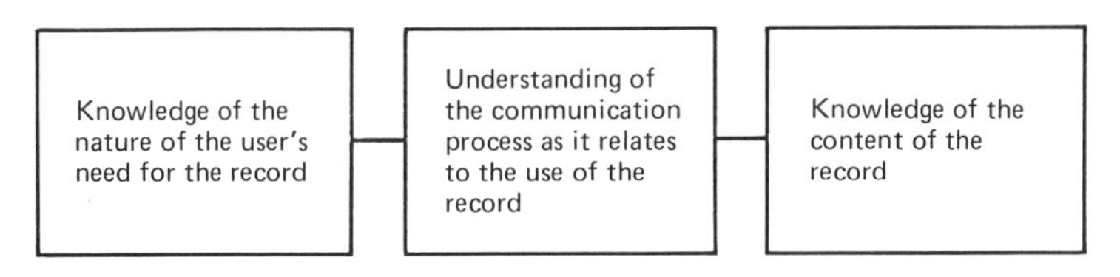

Figure 1.2 Three kinds of knowledge needed for effective management of records.

a subject level and a metalevel of operation in dealing with records (Rawski 1969). The *subject level* pertains to the ideas, concepts, and terminology of the subject field; the *metalevel* pertains to the formal structure and processes of that field's records, including such aspects as their physical format, their language, and their bibliographical apparatus.

An example may help clarify the distinction: A librarian with the requisite knowledge of physics may work with a physicist in the early definition of a problem in physics, performing many of the same conceptual operations that the physicist performs at that stage. When the librarian turns to the problem of searching for the related literature, it is likely he will shift to the metalevel, devising and using techniques based on his knowledge not of physics but of publication formats, prevailing linguistic patterns, and networks of authorship and publication. At this stage he is contributing a different element, one that is proper to his role as records manager.

The librarian, in acquiring and organizing subject literatures, is guided by whatever knowledge he has at both the subject level and the metalevel, but his unique contribution as a librarian comes from his knowledge of the metalevel, which is concerned with the literature rather than with the ideas it contains. Thus, the term *literature* implies (1) a body of ideas, (2) their expression in published form, and (3) the records themselves—the physical correlate. It is the third of these that constitutes the subject of the present work. The literature under consideration is studied as to its patterns of authorship, publication, and citation. These aspects may be referred to as *literature statics*, because they reflect the production and use of the literature at a given point in time. Other features are analyzed to determine changes in the literature over time, an area of analysis referred to here as *literature dynamics.*

The formal structure of literatures has long been the subject of bibliographic study. Bibliographers have traditionally employed formal elements to distinguish among types of publications. They have described publications on the basis of the national or trade systems that produce them, and they have related the development of these systems of publication to social and economic trends. This has contributed to our knowledge of literatures at given times and places, that is, to literature *statics.* Similarly, an illustration of the study of literature *dynamics* may be found in the fields of *belles lettres* and history, where researchers have traced the influence of publications by means of citation and use studies. For example, *The Road to Xanadu* describes the effect of Coleridge's reading upon his poetry (Lowes 1927), and scholars of eighteenth-century political history have traced the influences of writers on one another through the analysis of their respective patterns of diction.

But considerably less attention has been paid to the explication of *general principles*, that could apply to subject literatures independently of their special contents, that is, descriptive laws of bibliography. Progress toward a science of librarianship seems to require such laws and principles as well as a metric adequate to the measurement of bibliographic data. Robert A. Fairthorne has pointed out that the field of bibliographic analysis is only now beginning to develop a "numeracy" (as a complement to literacy), equal to the tasks at hand. Further, the concern of the mathematically competent investigators of documentary problems has been, until recently, quite restricted in scope. "Up to the late 1950s," says Fairthorne, "the most pressing problems concerned indexing vocabularies and their coding into forms congenial to various devices. All these ingredients were regarded as commercial commodities. Thus, though the growth of indexing vocabularies with the size of the collection came in for much attention, especially after 1950, studies of it were mainly *ad hoc* and confined to one's own system" (Fairthorne 1969, p. 338).

The 1950s were a time of concern with the problems of *science information*. Many scientists and documentalists at that time undertook to solve the difficulties that they believed attendant on the rapid rise in scientific publication. Some believed that the solution would be found in centralized, computer-aided documentation systems. Both panaceas, centralization and computerization, have since lost much of their magic appeal as the problems they entail have become better understood and as the study of documentary communication has been broadened to include the interaction of formal and informal information channels (Price 1961, 1963; Paisley 1968). The centralized computerized document retrieval service that once seemed the ultimate boon appears less attractive when it is understood that the mass of its offerings are irrelevant to the needs of any given set of users and that the use of the literature is variously affected by the patterns of informal communication that obtain within different groups. The physicist Ziman (1969, p. 322), in a critique of mechanized information retrieval, alludes to some of the behavioral and economic aspects of information use that have been ignored by many proponents of big mechanized systems. He comments

Much effort, for example, has been expended on the idea that everybody really has been waiting for a daily alerting service that will tell each one of us about the latest papers in the topics which interest us. Thus, I shall have given instructions, on a punch card, to be kept up to date on "the theory of liquids," on "electron theory of metals," on the "structure of alloys," and on "lattice conduction in solids;" hey presto, the latest abstracts or papers on just these topics will be waiting on my desk each morning by courtesy of Electronic Alerting Service for Yokels (EASY). But I don't really care that much about what everybody is doing in all these subjects; and anyway I have not the time to read through all this stuff carefully; and also I've got interested in spin waves again recently and have forgotten to inform EASY about this; and what I suddenly discovered last week was a paper published ten years ago which gave the solution to my problem in another form; and anyway I'm damned if I'm going to become the slave to this sort of guff, which is worse than the leaflets hopefully distributed by publishers; and I don't see why I should pay £100 a year for it out of my own pocket, when that would improve the lighting in the library, buy us decent blackboards, provide free tea for the whole research group, and could generally be spent more profitably. You see what I mean.

The reliance on automation and centralization seems to be based on the assumption that a given reader is likely to find any document in the store equally relevant to his information needs. Librarians and other information specialists know this isn't so, but until recently they have had no very exact way to predict which segments will be useful. The information-scientific methods explored in the present work offer promise of providing this predictive power.

Earlier, it was stated that the librarian's role as records manager involves three kinds of knowledge: knowledge of the record itself, knowledge of the user's need for the record, and knowledge of the process by which the user makes effective contact with the record. Obviously, the three are interrelated, but different kinds of methods are likely to be useful in investigating each. The nature of the user's need may require an investigation of his relation to his intellectual, social, and economic environment. The study of the process by which he makes effective contact with the record may require an analysis of his searching and learning habits and those of the librarian, as discussed by G. L. Gardiner (1969). But these lie outside the present study, except insofar as they may be elucidated by a study of the record itself. The kinds of library problems that may be solved by the techniques used here are those that have to do with the selection and organization of library materials.

In Chapter 2 certain information-scientific techniques are introduced that have been developed in connection with the study of particular literatures. In each case, the theory is discussed in the context of the original work, followed by some indications of the technique's potential value for libraries. In Chapter 3 these techniques are integrated into a general method that is then applied to a single literature. In Chapter 4 the results are analyzed with respect to what they show about the nature of the literature being examined. These findings are then explored for indications of their value in the solution of library problems.

2
Techniques

Certain information-scientific techniques for literature analysis are introduced in this chapter in the contexts in which they were first developed, and some relationships among them are shown. These include the Bradford distribution, citation tracing, bibliographic coupling, and the epidemic analysis of literature.

2.1 The Bradford Distribution

The *Bradford distribution* is a phenomenon found in the dispersion over a set of journals of articles that relate to a given subject. It was discovered by Samuel Clement Bradford, a distinguished British librarian and documentalist, an information scientist before the title was invented. In his role as Keeper of the Science Library, Bradford studied the "usefulness" of certain journals in fields of interest to his clientele. This study led to the discovery of the regularity later found to obtain in many scientific literatures, which Bradford termed a "law of scattering" and which is here called the Bradford distribution.

Bradford first published his law in an article entitled, "Sources of Information on Specific Subjects," in the January 2, 1934 issue of *Engineering* (London), which article constitutes a chapter in his book *Documentation*, published in 1948. The classic statement of the discovery is as follows (Bradford 1934):

> If scientific journals are arranged in order of decreasing productivity of articles on a given subject, they may be divided into a nucleus of periodicals more particularly devoted to the subject and several groups or zones containing the same number of articles as the nucleus, when the numbers of periodicals in the nucleus and succeeding zones will be as in $1{:}n{:}n^2$.

As a librarian, Bradford was interested in finding out which journals were most important for certain fields of study. Some journals published a great number of articles on a subject, others only a few. He arranged the journals into a list with the most productive numerically (in terms of the particular subject) at the top, followed by the next most productive, and so on, to the least productive, reasoning that those at the top would form a nucleus of the most useful sources for that area of study. In doing so, Bradford discovered an interesting fact. There appeared to be a regular relationship between the numerical productivity of journals in the nucleus and the productivity of journals in the zones that followed. For example, if the nucleus consisted of 200 papers appearing in four journals, and the next 200 papers appeared in six journals (1 1/2 times the number of journals in the first zone), then the following 200 would appear in nine journals (1 1/2 times the number of journals in the second zone). This regularity held for several different literatures. Thus, for each of these literatures, there was a constant, not always the same, called here the *Bradford multiplier* b_m, which governed the relationship between the number of journals producing the nucleus and the number in each succeeding zone. The constant varied for each literature, depending on the number of zones into which the list was divided. As the number of zones increases, the constant decreases, although it always remains greater than unity.

The statistical distribution that obtains here (the Bradford distribution) is discussed at length by Fairthorne (1969), who describes it as one of a family of empirical hyperbolic distributions, that is, those in which the product of fixed powers of the variables is constant. This family includes, among others, Zipf's first law, which gives the relation between the rank of a word in order of frequency and its frequency in a text of sufficient length. Zipf's law is discussed later as it relates to the general method being developed in this study.

Bradford demonstrated an empirical phenomenon but did not provide a formal mathematical statement of it. This has since been done by B. C. Vickery (1948), by F. F. Leimkuhler (1967), and by R. A. Fairthorne (1969). And although Bradford had recognized the need for determining the *minimal* nucleus (the smallest nucleus for which the law holds), he did not provide a mechanism for doing so. W. Goffman and K. S. Warren (1969) have shown that for a given literature the minimal nucleus exists and that it may contain more than only the most productive journal. In a study of the dispersion among journals of two quite unrelated medical literatures, they verified the applicability of Bradford's method. In a further extension of the method, Goffman and Warren showed that the ratio of articles to authors approximated the minimal Bradford multiplier, that is, that constant governing the subdivision of periodicals into the minimal nucleus and succeeding zones. Examining the bibliographies of ten medical researchers with respect to the dispersion of their work over journals, they found in all cases indications that these regularities occurred also in the scatter of individual bibliographies, with the average paper/author ratio approximately equal to the minimal Bradford multiplier b_m. This last regularity, appearing as it has in a number of subsequent studies including the present work, indicates that the average number of articles per author may be taken as a convenient trial value for the Bradford multiplier in a given corpus.

In the studies by Bradford, decisions as to the relevance of a given article to the field of interest were made by inspection. Once this was done, the statistical regularity described by his law provided an objective means of determining zones of relative richness or value to a given kind of library collection. This suggests applications to the library acquisition process. At present, libraries acquire materials on a pragmatic and largely intuitive basis, using so-called "standard lists" compiled by subject specialists, professional organizations, or vendors. But little has been done to provide more objective criteria. In a library serving a clientele that is oriented to problems or missions rather than to disciplines the choice of materials is often extremely difficult to make. The usual process is to identify those disciplines that seem to be relevant to the mission and to choose certain "leading" periodicals from within each subject specialty making up those disciplines. Studies are sometimes conducted of the use made of library collections based on introspective judgments of the users concerning current or past experience with the literature. Several dangers are apparent in this approach. One is that the user's response will be invalid as a result of poor memory. Another is that the respondent will bias his statement about the journals he uses toward the prestigious publications, feeling that these are the ones he is expected to read. A more critical problem, considering the rapid rate of change in contemporary research, is that past and current experience is an insufficient guide to future areas of interest. The interdisciplinary nature of scientific work and the unstable character of research funding in this era often cause research groups to focus their attention on a series of subjects in rapid succession. This seems to be particularly true of groups working on applications and engineering development. However, obsolescence in a collection as a result of changing client interests is not the only consideration; even under more stable conditions, economy demands careful selection of library materials in order to obtain the maximum benefit from the cumulatively large investment made in library development and maintenance.

A trivial solution to the selection problem would be to obtain all of the periodicals published, or alternatively, to obtain all of those in a given subject area defined. The first of these solutions is an impossibility and the second an absurdity in economic terms. It is apparent that there exists a point at which it is preferable to obtain a copy of an article upon demand, rather than to subscribe to the publishing journal on a permanent basis. Present trends in the direction of interlibrary cooperation, as well as improvements in the techniques for making and transmitting copies, make this point very clear. We may say that, given a librarian's desire to provide journals of most likely interest to users in a given subject area, the library should stock those journals that fall into the first zone, or minimal nucleus. If the library's budget permits, it may be desirable to extend the purchase list to include journals in succeeding zones. But how can the librarian determine the optimal point at which to stop adding journals?

Although the technique developed by Bradford provides a means for establishing a lower limit, it does not provide for an upper limit in periodicals acquisition. The lower limit could be used in a small library to determine which journals are most needed. But what of the problem of the large library, where the choice might be made to purchase all relevant journals or, alternatively, to buy only those of a certain degree of likely usefulness? How can a cutoff point be determined between those of high and those of low potential value?

This question is critically important for those large libraries that share resources and responsibilities on a national or regional basis. B. C. Brookes (1969) points out that the Bradford distribution can be used to estimate the total size of a bibliography and to guide the selection of fewer than the total number of contributing journals:

> If the total expenditure on periodical provision is limited to the fraction f of the sum needed to cover the subject completely, the buying of periodicals may be supplemented by the buying of photocopies of the relatively few relevant papers published in the peripheral periodicals.

In a national system, with respect to specialized needs, this could work to reduce the load borne by the individual library in the following way, as stated by Brookes (1969, p. 956:)

> The law could also be applied to the planning of special library systems; the model considered here consists of a national, a regional, and a local library. The local special library is assumed to be interested in a scientific subject whose complete periodical bibliography consists of N periodicals with the characteristic constant s. Of these N periodicals, the national library takes all, the regional library takes n_1, and the local library takes n_2, where $c \leqslant n_2 \leqslant n_1 \leqslant N$. Most of the periodicals taken by the regional and national libraries will be in the outer Bradford zones of the subject and therefore contain papers of interest to other special libraries as well.

It is not necessary, however, to assume that the national or central library in a system would acquire all of the journals. We might assume, rather, a cooperative system that would assure the acquisition of every desired journal by at least one member library, which could then provide copies to the other libraries upon request. However, a member library must still distinguish between those journals of primary value to its users and those of lesser importance. Discoveries by Zipf (1935) and Booth (1967) led Goffman to suggest a method for determining a maximal limit for acquisitions (private communication, 1969).

Zipf's first law was developed in relation to frequency of use of given words in a text of sufficient length. Words were ranked from high to low frequency of occurrence in the text, that is, from common to rare. The formula states

$$rf = c$$

where r is the rank and f is the frequency of occurrence in the text and c is a constant for a given text. As in the case of the bibliographies studied by Bradford, the dispersion was found to be hyperbolic, holding for the high-frequency word occurrences. Zipf's second law, as restated in more general form by Booth (1967) relates to low-frequency word occurences and is stated as

$$\frac{I_1}{I_n} = \frac{n(n + 1)}{2}$$

where I_n is the number or proportion of distinct words each of which appears in the text exactly n times, and I_1 is the number or proportion of distinct words each of which occurs only once.

Booth found for rare (low-frequency) items a hyperbolic distribution that in continuous terms can be expressed as

$$r(1 + f) = c$$

or, for discrete values, may be put in the form

I_n = number of words that occur n times in the text

= the rank of words that occur $(n + 1)$ times less the rank of those that occur n times

$$= \frac{2I_1}{n(n + 1)}.$$

Hence,

$$\frac{I_1}{I_n} = \tfrac{1}{2}n(n + 1)$$

where I_n is the number or proportion of distinct words that each appears in the text exactly n times, and I_1 is the number or proportion of distinct words each of which occurs only once.

Thus, Zipf's first law for common words and Booth's law for rare words in the same text are not the same, but we can find a transition value of n where both laws give the same result, so that Zipf can be applied to common words and Booth to rarer words without discontinuity. Goffman has suggested that this value can be predicted as the value of T given by

$$T = \frac{-1 + \sqrt{1 + 8I_1}}{2}$$

Then T is a point of transition between high and low frequency words.

This suggests that T, as applied to the journal dispersion, might be used to partition the distribution into high and low frequency journals. Just as the minimal nucleus provides a lower limit of periodicals, then T provides the upper limit. This point of transition can be taken by the library as a reasonable cutoff point in acquiring periodicals relating to a given subject.

As Brookes (1969, p. 954) indicates

The Bradford-Zipf distribution can be expected to arise when selection is made of items, characterized by some common element, which are all equally open to selection for an equal period and subject to the "success-breeds-success" mechanism, but when the selection of a most popular group is also, but to a weaker extent, subject to restriction. It is thus a general law of concentration over an unrestricted range of items on which is superimposed a weaker law of dispersion over a restricted range of the most frequently selected items.

Thus, the Bradford-Zipf distribution might be expected to apply not only to the dispersion of articles over journals and to the frequency of words in a text but also to such phenomena as the distribution of users, or of books circulated in a library, over the subject matter fields served, and so on. In fact, Goffman and Morris (1970) have demonstrated its applicability to these cases, as well as to the distribution, by diseases, of patients in a medical clinic.

Bradford's method promises to be a useful tool in increasing the effectiveness of library acquisition procedures. Acknowledging the need for further verification of its applicability to bibliographic work and of the reliability of estimates derived from its application, Brookes (1969, p. 956) concludes

It seems to offer the only means discernible at present of reducing the present quantitative untidiness of scientific documentation, information systems, and library services to a more orderly state of affairs capable of being rationally and economically planned and organized.

2.2 Citation Tracing and the Research Front

The Bradford technique gives us a means of identifying journals that are particularly relevant to a subject. Within any such set of journals at a given time, there will be certain articles that contain the most seminal concepts in the development of the discipline, and it is important for the librarian to be able to identify those contributions. Indeed, the improvement of library services appears to demand that he be able to do so. With respect to the identification of this body of seminal literature,

Price (1965) has evolved the notion of the *research front*. A review of his work will be helpful in understanding this concept.

After examining several large bibliographic data bases, Price (1965, p. 511) found that papers in journals contain an average of 15 references per paper, most of which are to other articles in journals rather than to books, theses, or other types of sources. Each year there are about 7 new papers for every 100 previously published in a given field. An average of about 15 references in each of these 7 new papers will therefore supply about 105 references back to the previous 100 papers, which will therefore be cited an average of a little more than one each during the year.

The analysis of citations by Price yielded the data found in Table 2.1.

Price found that

> The maximum likely number of citations to a paper in a year is smaller by about an order of magnitude than the maximum likely number of references in the citing papers. There is, however, some parallelism in the findings that some 5 percent of all papers appear to be review papers, with many (25 or more) references, and some 4 percent of all papers appear to be "classics," cited four or more times in a year.

Table 2.1. Citation Analysis (after Price)

Percent of Papers	Times Cited (average)
16	3.2
9	2
3	3
1	5
1	⩾ 6

Some papers are never cited, and others are cited only for a few years. Price believes that for the latter type of paper the chance of being cited at least once in a given year is about 60 percent. The useful life of an article, judged by references to it, appears to be about ten years. He concludes that extended study of citation patterns could lead to the discovery of classic and "superclassic" papers that could be picked automatically by means of citation index procedures and published as a single international publication, a *Journal of Really Important Papers*. Study on a smaller scale indicates that the overall pattern is highly selective, with each group of new papers related to a small portion of the existing scientific literature but with weak and random connections to a much larger part. "Since only a small part of the earlier literature is knitted together by the new year's crop of papers, we may look upon this small part as a sort of growing tip or epidermal layer, an active research front" (Price 1965, p. 512). In his view, the scientist working in a field whose literature evidences a research front needs some kind of current alerting service, such as a citation index. Other disciplines, which Price calls "taxonomic," treat every previously published work as being of permanent interest and organize their literatures by more or less fixed classification schemes. For literatures that are not clearly either research front or taxonomic, he believes citation study can establish a conceptual map (Price 1965, p. 515):

The present discussion suggests that most papers, through citations, are knit together rather tightly. The total research front of science has never, however, been a single row of knitting. It is, instead, divided by dropped stitches into quite small segments and strips. From a study of the citations of journals by journals I come to the conclusion that most of these strips correspond to the work of, at most, a few hundred men at any one time. Such strips represent objectively defined subjects whose description may vary materially from year to year but which remain otherwise an intellectual whole. If one would work out the nature of such strips, it might lead to a method for delineating the topography of current scientific literature. With such a topography established, one could perhaps indicate the overlap and relative importance of journals and, indeed, of countries, authors, or individual papers by the place they occupied within the map, and by their degree of strategic centralness within a given strip.

The conceptual map that emerges from such study can be the foundation for a classification scheme that would be dynamic rather than fixed, since it could reflect changes in the conceptual relations involved. This classification scheme could then be used by the library in the management of publications.

In Chapter 3 citation tracing is used to show links among papers in a particular literature.

2.3 Bibliographic Coupling

The research front for a given subject area may be very broad, capable of being subdivided into smaller subfronts. A general method of classifying a file according to such subfronts, in terms of the relatedness among members of the file, is given by Goffman (1969) . He objects to the treatment of the information retrieval process as strictly a matching procedure where every element in a file is examined with respect to some subject to determine the relevance relationship. This, he indicates, overlooks the fact that what is learned from one document affects what may be learned from subsequent documents examined. He presents a mathematical method of search strategy in which this is taken into account. The results of an experiment conducted along the lines of this model indicate that Goffman's method, which he calls an *indirect method* of information retrieval, obtains better results than the direct matching procedures. The method of classification that results as a by-product of the indirect method is as follows (Goffman, private communication 1969) :

Given a collection of documents X, then p_{ij} is defined as a conditional probability that x_j in the set X is relevant to some query if x_i is relevant to it. Thus p_{ij} represents the conditional relevance measure for x_j, relative to x_i, and in a sense represents a measure of subject relatedness. A *communication chain* is defined as a sequence of elements x_i to x_j such that the conditional relevance measure $p_{k-1,k}$ exceeds some critical probability, where $k = i + 1, \ldots j$. Finally, two elements x_i and x_j are said to be in *intercommunication* if there exists a communication chain from x_i to x_j and vice versa. Intercommunication turns out to be an equivalence relation. Hence the collection X can be subdivided into intercommunication classes on the basis of the interrelatedness of the information contained in the documents in X. Such a partition constitutes for the members of the collection, a classification that can be made either fine or coarse, depending on the selection of the critical measure. The use of a critical measure of zero would produce the most general classes whereas a number very close to one would produce the most specific.

The literature of information retrieval abounds with special schemes (the so-called statistical association measures, Stevens 1965) , any one of which can be used in an experimental setting as a means of estimating the conditional measures p_{ij} as stated in this general theory. In a simple and useful method developed by M. M. Kessler, a single item of reference used by two papers is defined as a unit of coupling between them. Kessler (1963, p.10) uses two graded criteria of coupling:

Criterion A: A number of papers constitute a related group G_a if each member of the group has at least one coupling unit to a given test paper P_0 . The coupling strength between P_0 and any member of G_a is measured by the number of coupling units n between them. G_a^n is that portion of G_a that is linked to P_0 through n coupling units. Criterion B: A number of papers constitute a related group G_b if each member of the group has at least one coupling unit to every other member of the group.

The coupling thus provides a clue to the relatedness of papers on purely structural rather than substantive grounds. The seasoned maker or user of library catalogs and bibliographies already employs a great number of similarly formal clues in establishing the relatedness of separate publications, including a complex network of associations that may send him in search of an author associated with a given subject, a publisher known to have specialized in a given field, or a university where related research is known to have been done.

In classification, as in retrieval, use of such external clues is in some respects superior to the actual examination of the text by the classifier or searcher. The linguistic problems of classification on the basis of text have been dealt with in depth by Borko (1967) . By contrast, Kessler has pointed out that the use of citation coupling for classification is independent of language. This avoids the difficulties attendant on vocabulary, syntax, and usage; in fact, the text need not even be examined. In addition, the classification scheme thus derived need not be fixed but can change in order to reflect changes in the collection, in the disciplines it represents, or in the nature of the users' needs.

In Chapter 3, bibliographic coupling is used in creating a classification for the research front of a particular literature. The method might be extended to the classic problems of library classification.

2.4 Epidemic Theory

The principal approach used in this study for the analysis of literature dynamics is the epidemic theory of literature growth, as developed by Goffman (1964) . Its value for the study of literature derives from the analogy between the spread of infectious materials and the spread of ideas. There are well-developed mathematical models used in medicine for the prediction of epidemics, that is, conditions where the incidence of a disease is clearly in excess of normal expectations. These methods involve the recording of the rate at which infectives accrue with respect to time and the extrapolation of the resulting epidemic curve. If this rate of change is positive, the process is said to be in an epidemic state; if negative, the process is in a decreasing state; and if constant, in a state of stability. The *size* of the epidemic is the number of infectives occurring during the course of its development. Its *intensity* is the ratio of its size to the total population in which it develops. The mathematical model used to represent an epidemic process may be either deterministic or stochastic. The deterministic model represents the process as a system of differential equations, while the stochastic model describes it as a finite state Markov process. In the application of the theory to a large population, the deterministic approach is adequate.

The usefulness of such mathematical models depends upon being able to solve the model and upon obtaining reliable measurements in order to test it. The elements involved are (1) a population and (2) an exposure to infectious material. The population consists of three types of members: infectives I, who have the disease; susceptibles S, who, given effective contact, are able to contract the disease and thereby to become infectives; and removals R, who for any reason cannot become infectives. Disease can be spread directly from person to person, that is, from I to S. It can also be spread indirectly by way of an intermediate host, sometimes called a vector. It is this latter case that is most clearly analogous to the spread of ideas through the literature. The correspondence of elements in these two situations, namely, the spread of disease and the spread of ideas, may be seen in Table 2.2. The similarities between the two conditions are acknowledged in common parlance when we speak of "the germ of an idea." An idea, analogously to a disease germ or a plant germ, has the potential for growth and reproduction, and of course mutation. Certain philosophic, artistic, and scientific ideas appear to have followed an epidemic process: They were held by isolated individuals in the population for a time, then became epidemic, and finally either became stabilized or declined.

Table 2.2. Elements of the Epidemic Process (after Goffman)

Elements of the Epidemic Process	Element as Interpreted in terms of:	
	Disease Epidemic	Idea Epidemic
Infective	Person with disease	Author with idea
Susceptible	Person who may get the disease	Reader who receives the idea
Infectious Material	Disease germs	Ideas

Goffman (1966) has applied epidemic theory to a comprehensive bibliography of mast cell research that included all aspects of the subject from the discovery of the mast cell in 1877 to 1963 (Selye 1965) . In order to establish an epidemic curve, it was necessary to assign dates of *infection* and *removal* for each author with respect to his membership in the population of researchers in this field. Because it is not feasible to determine the time at which an author begins or ceases thinking about a subject, an author was considered to have become infective during the year of publication of his first paper cited in the Selye bibliography and a removal one year after the date of his last paper in that list. Rates of change were then computed for the population of contributors.

In graphing the curves, intervals of 5 years were used to remove the effect of fluctuations within smaller intervals, because these fluctuations were not considered significant in the overall time span of 86 years. Periods of stability and instability were noted, and a prediction was made that research on this subject would reach a peak in 1978, approximately 30 years after the beginning of an epidemic. In a further refinement of the method, Goffman analyzed the epidemic curves for three subject areas within mast cell research, each associated with a particular discovery. From this analysis he was able to demonstrate that at certain times there were shifts in emphasis among the aspects to which researchers directed their attention, and to indicate the size and direction of these shifts. Goffman has also demonstrated that for a given field there is a predictable point beyond which continued growth in its literature impedes the efficient flow of ideas.

It appears that there are many areas of decision making in libraries that could benefit from the application of epidemic analysis. It could be used, for example, to predict the need for a special library in a certain subject field or geographic region and could indicate how long such a library would be likely to be useful. Within a general subject area, a library could plan its collection more wisely by using these techniques to predict the emergence of activity in a subfield, as well as the duration, size, and intensity of that activity. By planning ahead for the time when a discipline's literature will become too large to support efficient communication, the library could make desired changes in its system for information retrieval and dissemination.

Such empirical evidence of well-defined needs must necessarily supplant vague statements of library policy if there are to be any substantial improvements in acquisitions and in library services more generally. In combination with other techniques described here, epidemic theory can be used to establish priorities of service and to choose among alternative means, through its power to predict changing patterns in literature production, publication, and use.

3
Analysis of a Literature

In this chapter, a particular literature is analyzed by each of the techniques discussed in Chapter 2, namely, (1) the Bradford distribution, (2) citation tracing to establish the research front, (3) bibliographic coupling and (4) epidemic analysis.

3.1 The Main Literature

The bibliography studied here is a subset of the literature of information science (IS) , an interdisciplinary subject devoted to the study of communication processes and information systems. This literature, called here the *main corpus* and denoted by M, is made up of all journal articles referenced in *Information Sciences—1967* (AFOSR 1968) . That publication is a summary of research activities supported by or administered by the Directorate of Information Sciences, Air Force Office of Scientific Research (AFOSR) , and the citations found in it are references to publications produced in connection with those research projects. The designation of these references as the primary data was made for two reasons. First, the bibliography is a result of an integrated program devoted to promoting theoretical reseach in information science. Second, it is a preselected set, which makes unnecessary any elaborate criteria for inclusion.

The reason for choosing only journal articles is that they are the contributions that have been subjected to review by the authors' peers and have been found acceptable for publication, whereas other kinds of literature, such as reports and contributions to proceedings, are often subject to little or no formal review. Thus, in its final form, the main corpus omitted the following types of items:

1. Those listed as "submitted for publication" or "in preparation;"

2. Papers delivered to meetings, convocations, symposia, or other gatherings but that did not also appear as journal articles;

3. Journal articles that consisted entirely of bibliographies;

4. Book reviews (state-of-the-art reviews that synthesized the works of several authors were included, however) ; and

5. Books.

Corpus *M* consisted of 160 articles *A* produced by 104 individual authors *a* and dispersed among 88 different periodicals *J* over a 10-year period from 1958 to 1967. In order to show the disciplinary areas of the articles in the main corpus, some system of classification was needed. The subject categories used are from *Ulrich's International Periodicals Directory* (Chicorel 1967) , a standard library reference work, and are listed here:

Abstracting and Indexing Services
Aeronautics
Anthropology
Archaeology
Architecture
Automation
  General
  Computers
  Data Processing and Operations Research
  Documentation
Banking and Finance
Biological Sciences

- General
- Biology
- Biophysics
- Microbiology
- Physiology
- Zoology

Chemistry
- General
- Analytic Chemistry
- Biological Chemistry
- Crystallography
- Electrochemistry

Communications
- Telephone and Telegraph

Education

Electricity and Magnetism

Engineering
- General
- Mechanical Engineering
- Chemical Engineering

General Periodicals
- United States

Instruments

Library Periodicals

Linguistics and Philology

Literary and Political Reviews

Management
- General
- Industrial Management

Mathematics

Medical Sciences
- General

Metallurgy
Pharmacy and Pharmacology
Philosophy
Physics
General
Nuclear Energy
Political Science
Psychology
Public Administration
Science
Sociology
Sound Recording and Reproduction
Statistics
Technology

The production of articles in corpus *M* according to Ulrich classes by year is shown in Table 3.1, with a breakdown of subcategories given in Table 3.2.

The production curve in two-year increments for articles in *M* is seen in Figure 3.1, and for articles in the seven most productive Ulrich classes in Figures 3.2 to 3.9. The overall production curve shows a geometric increase beginning in 1965, as do the curves for most of the fields separately charted. The one exception is Automation—Documentation, which began a decline in 1965.

3.1.1 Bradford Analysis of the *M* Corpus Journals

The procedure for the Bradford analysis of journals is as follows:

1. Tally the articles appearing in each journal.

Table 3.1. *M* Corpus Articles, Production by Year and Subject Class

Subject	58	59	60	61	62	63	64	65	66	67	Total
Abstracting and Indexing						1	2		1		4
Aeronautics										3	3
Automation			2			7	3	7	16	13	48
Banking									1		1
Biological Sciences								1	2	4	7
Chemistry									1		1
Education									1	1	2
Electricity and Magnetism				1		2	2	1	9	4	19
General Periodicals					1						1
Library Periodicals						3	1			1	5
Linguistics and Philology								1			1
Literary and Political			1							1	2
Management					1					1	2
Mathematics						2			8	5	15
Medical Sciences			1	3	1		1	1	4	2	13
Philosophy	2						1	1	5		9
Physics					1			1	3	2	7
Political Science										1	1
Psychology					1			1	1	2	5
Science					1	3	2		4	2	12
Sociology					1						1
Sound Recording										1	1
Totals	2	0	4	4	7	18	12	14	56	43	160

2. Arrange the journals in order of decreasing productivity.

3. Divide the list into zones, such that they contain the smallest equal number of articles that will effect the Bradford partition of the list.

4. Establish the ratio between the number of periodicals in the nucleus and the number in each succeeding zone. This is the Bradford multiplier b_m for journals in the main corpus.

The distribution of corpus *M* articles by journals is shown in Table 3.3. The above procedure, applied to the data, produced a Bradford distribution as shown in Table 3.4.

The five periodicals in zone 1 of Table 3.4 thus constitute the minimal nucleus or core of journals devoted to the subject represented by this literature. The journals are:

ACM Communications
Acoustical Society of America Journal
IEEE Transactions on Electronic Computers
Information and Control
Information Storage and Retrieval

These would be selected as the minimal collection of a library concerned with this literature. If it were desirable to have a

Table 3.2. Subcategories of Classes in Table 3.1

Subject	58	59	60	61	62	63	64	65	66	67	Total
Automation, General			2			2		4	2	3	13
Automation, Computers						2		1	11	8	22
Automation, Data Processing and Op. Res.									1	1	2
Automation, Document.						3	3	2	2	1	11
Biological Sciences, Biology									2	4	6
Biological Sciences, Biophysics								1			1
Chemistry, Analytic									1		1
General Periodicals, United States					1						1
Medical Sciences, General			1	3	1		1	1	4	2	13
Physics, General					1			1	3	2	7

larger collection, journals would be added from succeeding zones of order of productivity. To establish the optimal cutoff point for such acquisition of periodical titles, we would apply the method suggested by Goffman, as discussed in Section 2.1, finding that the transition point T is 10; journals contributing more than this number of articles to the bibliography would be considered to be of high productivity. Because there are no journals of high productivity for this literature, it follows that the minimal nucleus constitutes the optimal collection for M, denoted by N_M.

Table 3.3. *M* Corpus Journals—Distribution

J	*A*	Total
3	7	21
1	6	6
1	5	5
4	4	16
9	3	27
15	2	30
55	1	55
88		160

Table 3.4. Bradford Distribution of Journals in *M*

Zone	*J*	*A*	Ratio
1	5	32	
2	9	31	1.97
3	14	32	1.55
4	27	32	1.92
5	33	33	1.22
	88	160	mean =1.67

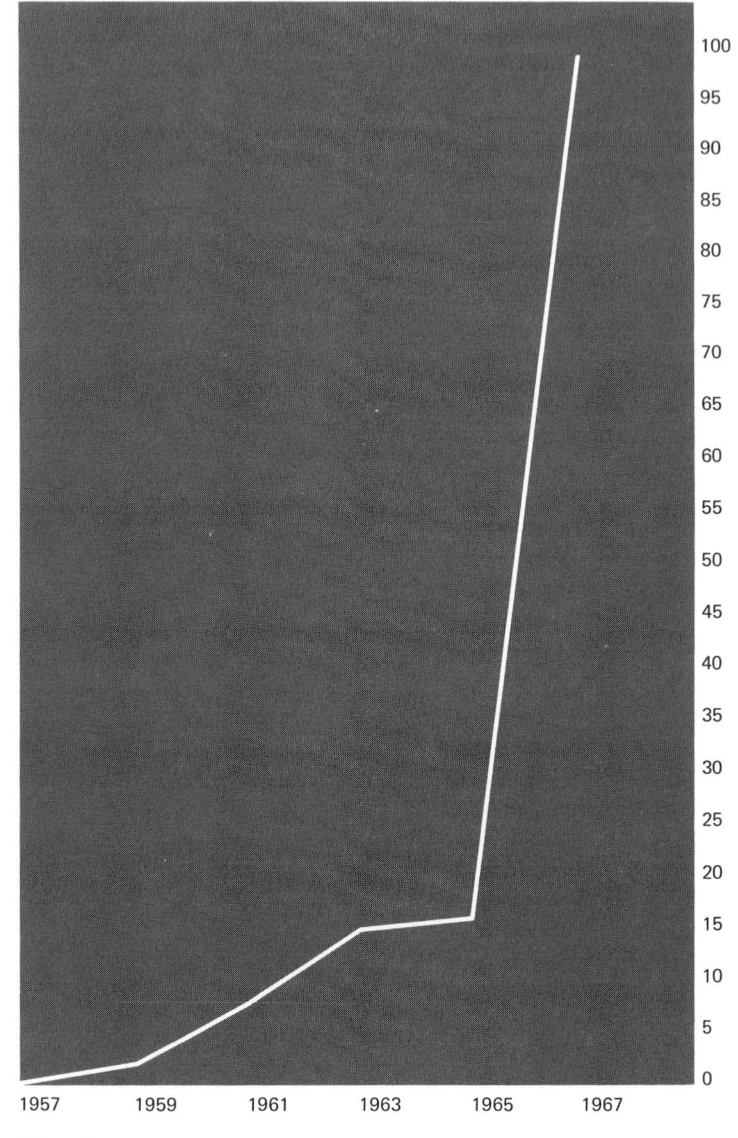

Figure 3.1 *M* Corpus articles

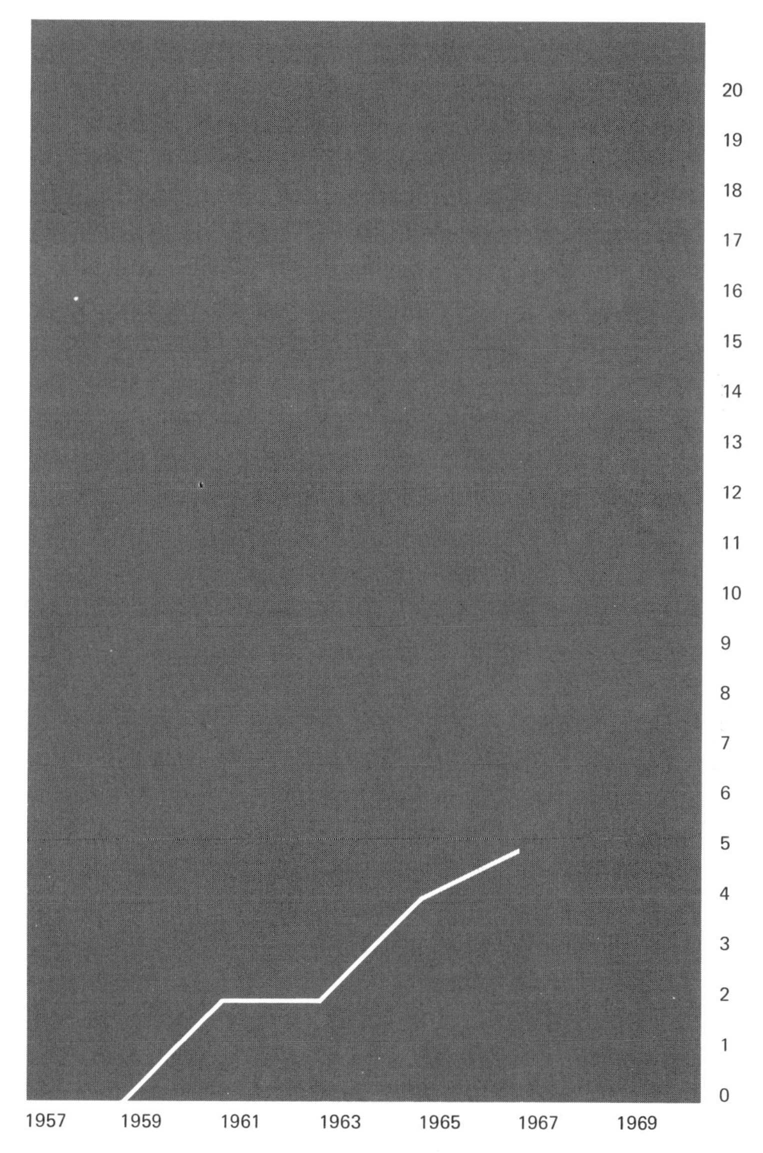

Figure 3.2 *M* Corpus—Automation, General

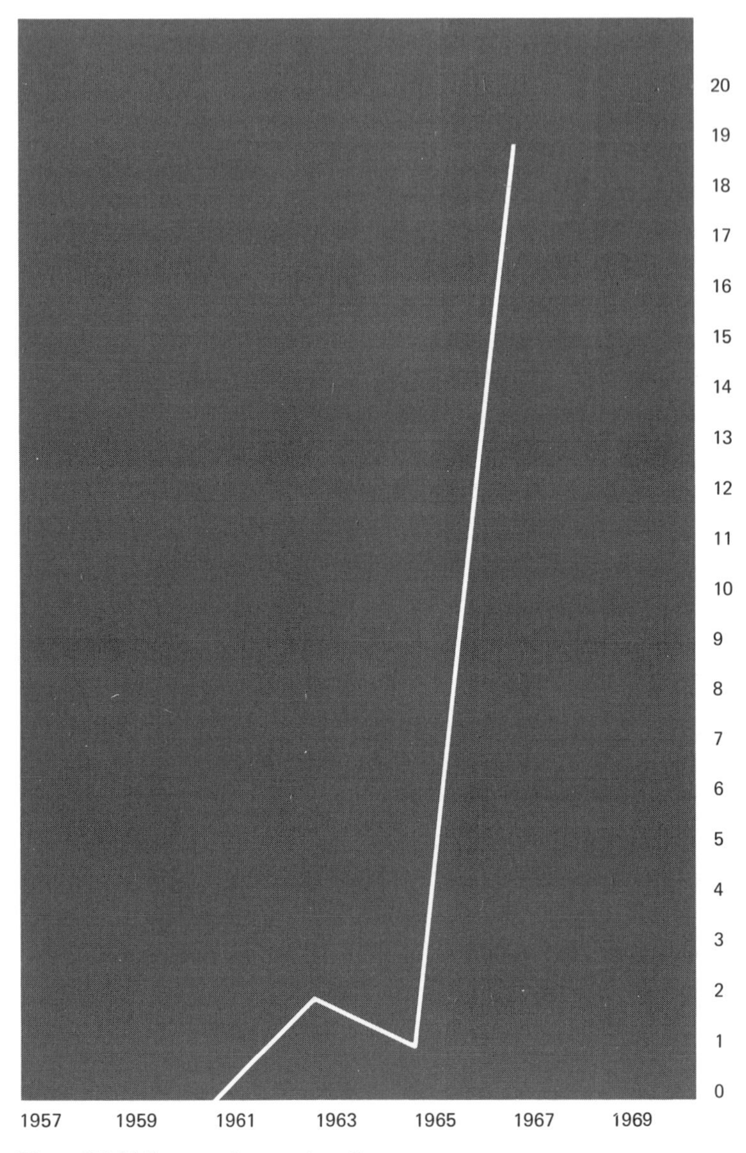

Figure 3.3 *M* Corpus—Automation, Computers

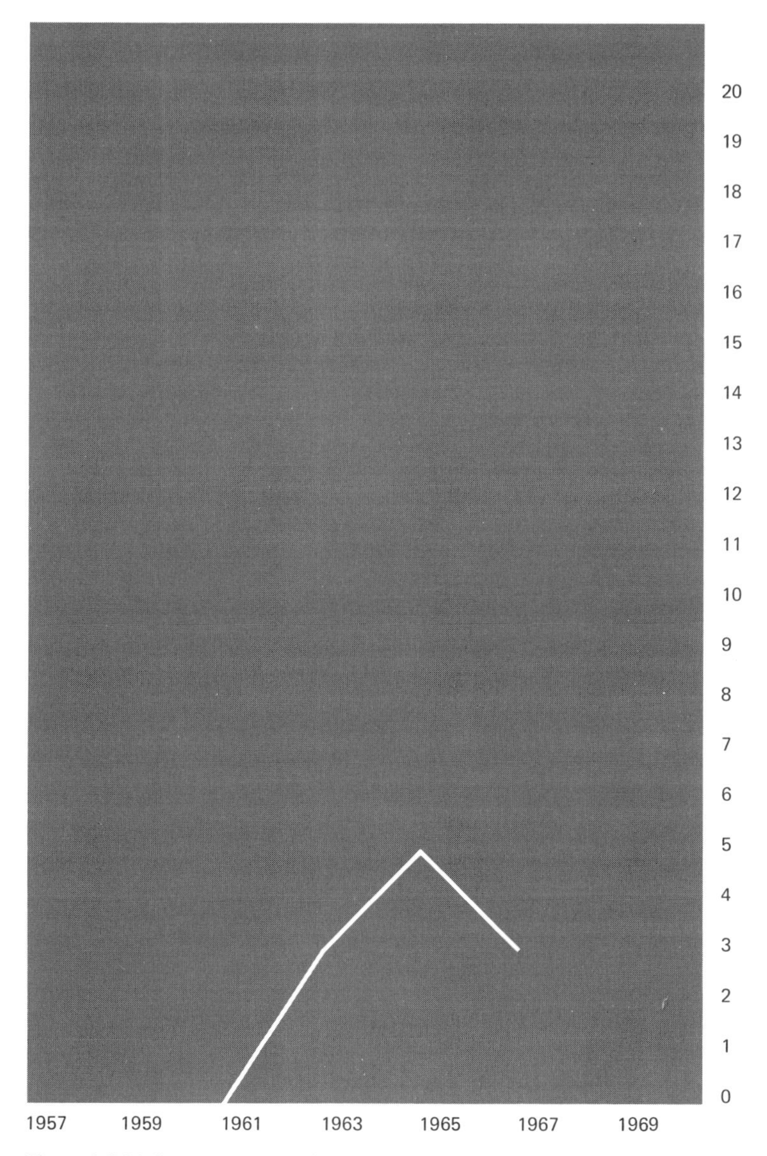

Figure 3.4 *M* Corpus—Automation, Documentation

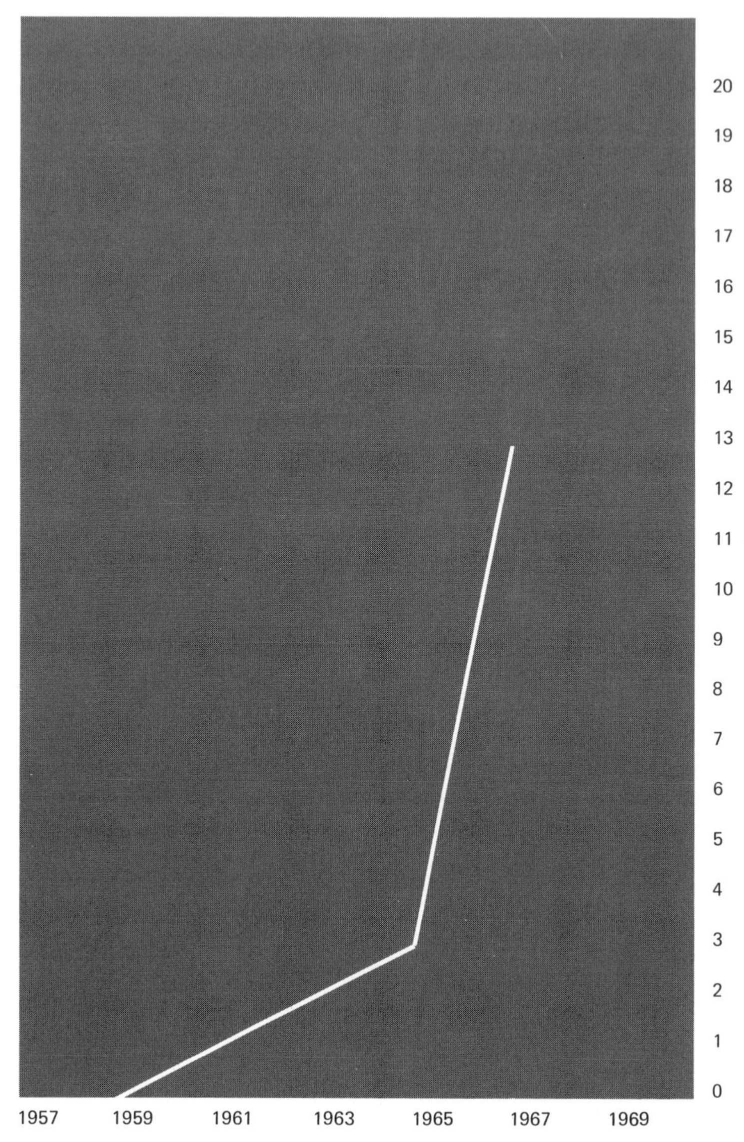

Figure 3.5 *M* Corpus—Elcctricity and Magnetism

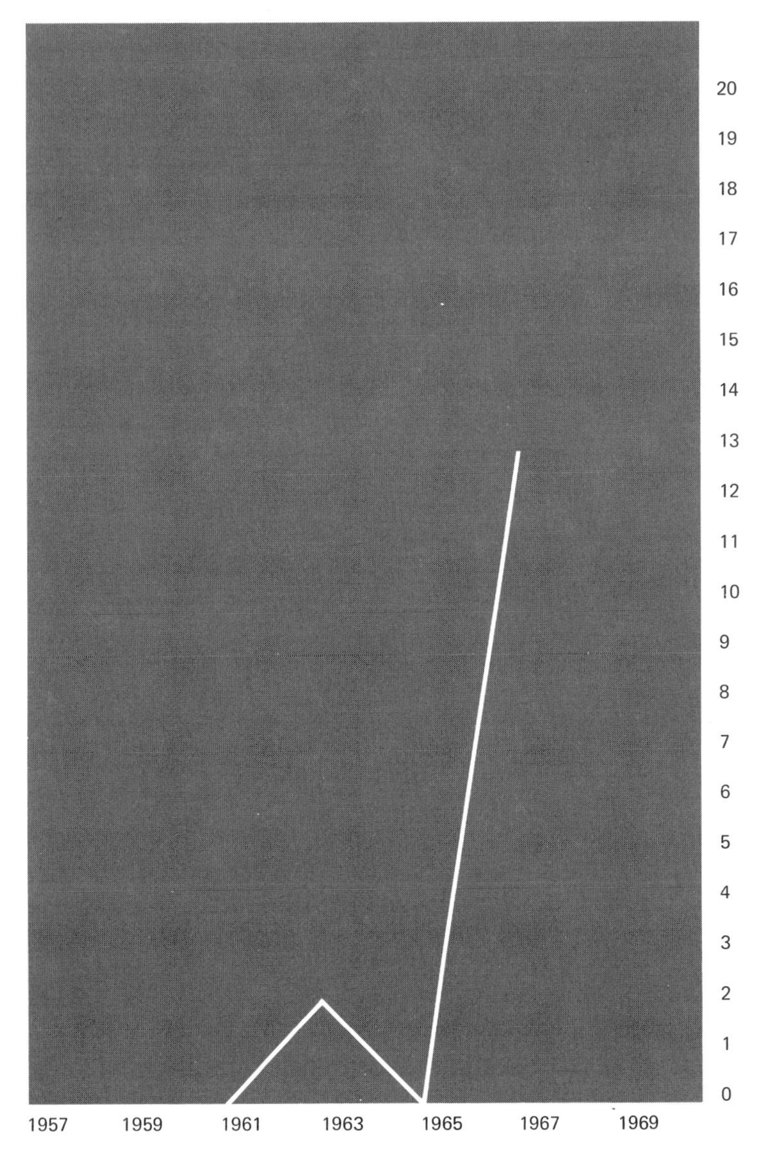

Figure 3.6 *M* Corpus—Mathematics

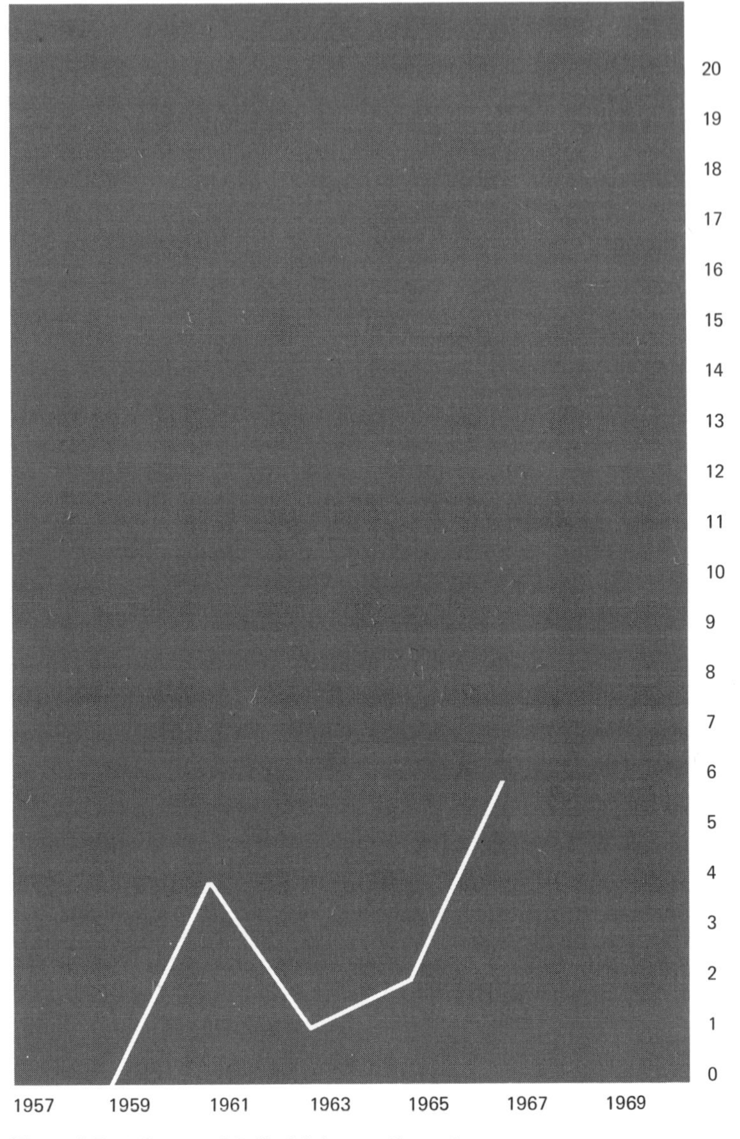

Figure 3.7 *M* Corpus—Medical Sciences, General

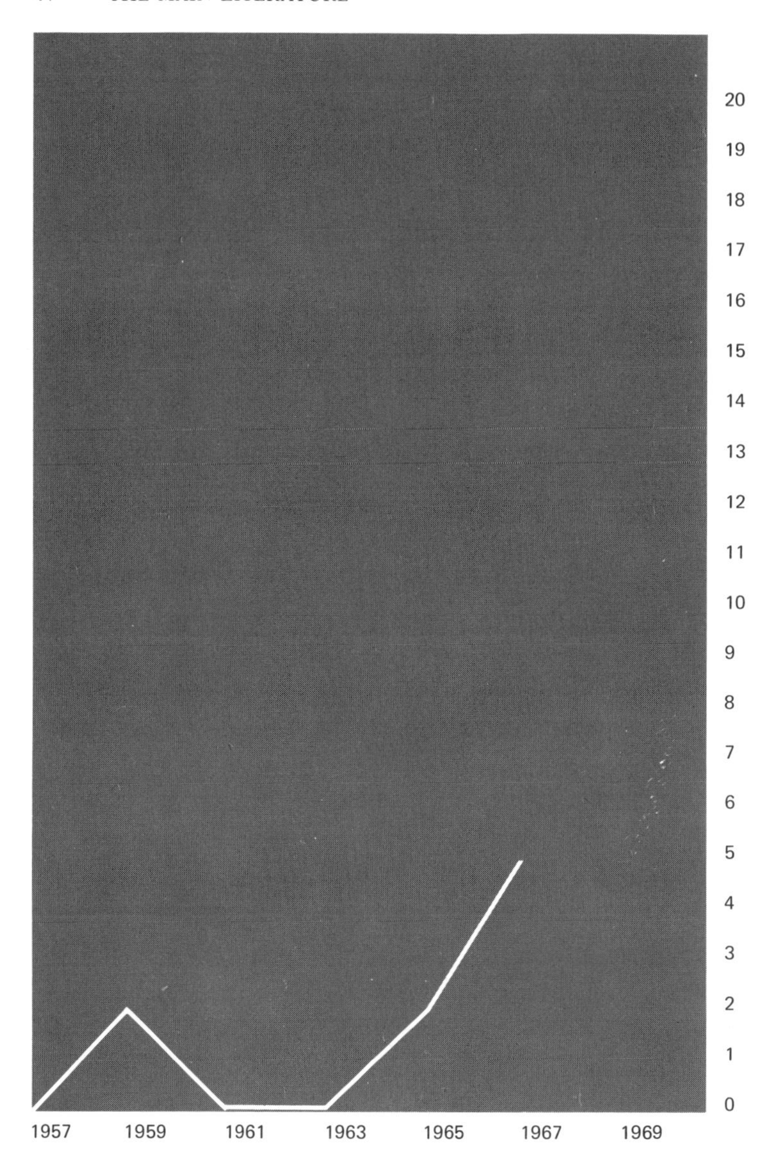

Figure 3.8 *M* Corpus—Philosophy

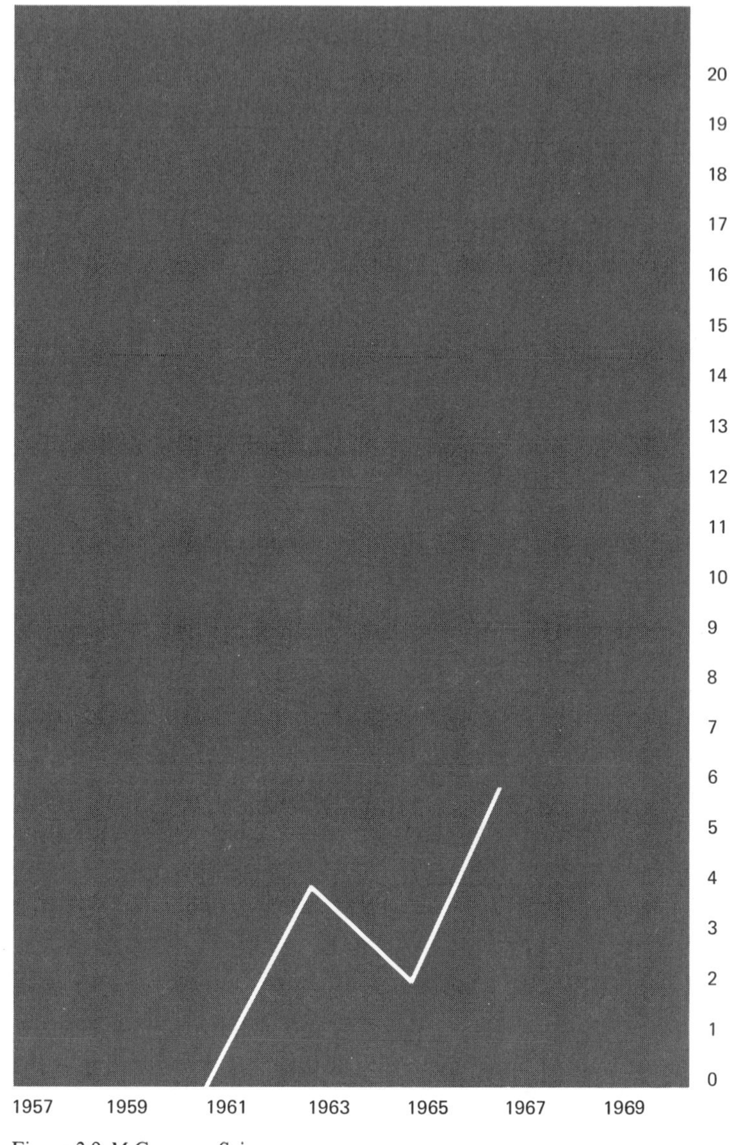

Figure 3.9 *M* Corpus—Science

The most productive journals in *M* are in the fields of physical science and engineering, including computing and the engineering aspects of information retrieval. This reflects the "hard science" orientation of the sponsoring agency. In the field of Automation, Documentation, we find included in the minimal nucleus the journal *Information Storage and Retrieval*, which is highly theoretical and mathematical, but not *American Documentation*, which is considerably more oriented to applications and to the description of library and document systems, either currently in operation or proposed.

3.1.2 Bradford Analysis of *M* Corpus Author Productivity

The productivity of authors in corpus *M* is shown in Table 3.5, followed by the Bradford distribution of authors in Table 3.6. As shown, the dispersion of articles over authors was Bradfordian, with the ratios between the number of authors in the nucleus and in succeeding zones ranging between 1.3 and 2. The proximity of b_m, the average minimal Bradford multiplier and g, the average number of articles per author, is striking,

Table 3.5. *M* Corpus Author Productivity

Number of Authors	Number of Papers	Total
2	11	22
1	10	10
1	8	8
1	6	6
3	5	15
6	4	24
10	3	30
14	2	28
66	1	66
104		209

Table 3.6. Bradford Distribution of *M* Corpus Author by Productivity

Zone	a	A	Ratio
1	3	32	
2	6	33	2.0
3	9	32	1.5
4	13	32	1.4
5	25	32	1.9
6	32	32	1.3
7	16	16	

$b_m = 1.6$
Average articles per author $g = 1.5$

being 1.6 and 1.5, respectively. Similarly, b_m in the analysis of the article dispersion over journals approached g, with a value of 1.67. These findings seem to verify the phenomenon noted by Goffman and Warren (1969), namely, that the minimal Bradford multiplier b_m converges to g, the paper/author ratio.

It seems remarkable that the minimal Bradford multiplier for the analysis of articles over authors and for the analysis of articles over journals should be the same. One possible explanation would be that the nucleus of authors accounts for the nucleus of periodicals by the volume of their publications. In this instance, however, such was not the case. The authors in the author nucleus did not publish with any great frequency in the journals that formed the journal nucleus. Therefore, it appears that the minimal nucleus of periodicals represents the output of the total population rather than simply that of the most active contributors.

3.1.3 The Research Front

Price's method for establishing the research front, as discussed in Chapter 2, was applied to the main literature. As a result, 32 of the 160 articles in the main corpus were found to belong to the research front. The technique of bibliographical coupling, as described by Kessler, was applied to the 32 members of R for the purpose of obtaining a measure of their content relatedness. Thus, for every two articles x_i and x_j in M, such a measure was defined as follows:

1. Establish the coupling strength between x_i and x_j, that is, the number of citations that these two articles have in common. Denote this coupling strength by $m(x_i \cap x_j)$.

2. Denote the number of citations to each of the articles x_i and x_j by $m(x_i)$ and $m(x_j)$, repectively.

3. The p_{ij} is defined as $m(x_i \cap x_j)/m(x_i)$ and p_{ji} as $m(x_i \cap x_j)/m(x_j)$.

By following this procedure for every two articles in the research front, we obtain a matrix of conditional relevance measures. Hence, we can subdivide R into disjunct intercommunication classes by the method of Goffman, as described in Chapter 2.

The taking of zero as the critical probability effected a partition of the research front into 13 intercommunication classes, each of which contains articles belonging to the same subject area. These are as follows:

1. Acoustics
2. Computing
3. Cybernetics
4. Documentation
5. Engineering
6. Information retrieval

7. Logic
   a. Recursive functions
   b. Reducibility
   c. Nested structures
8. Linguistics
9. Numerical mathematics
10. Statistics

There were three different classes dealing with special topics in Logic. Thus, the research front R of the main corpus M consisted of articles dealing with ten distinct disciplines. Therefore, even though this sample of the field of information science appeared to represent a wide variety of interdisciplinary activity according to the Ulrich classification scheme, its research front consists of 10 areas of activity. In fact, the research front is not a front properly speaking, but rather 10 disjunct fronts that taken together seem to make up the interdisciplinary field of information science as defined by the M literature.

The journals that were found earlier to be in the minimal nucleus of the Bradford distribution were

ACM Communications
Acoustical Society of America Journal
IEEE Transactions on Electronic Computers
Information and Control
Information Storage and Retrieval

It is interesting to note that these journals are in fact representative of disciplines belonging to the research front, implying that

1. The nucleus is an appropriate set of journals devoted to the field of information science as represented by corpus *M*, and

2. It is apparently the activity in the research front that established the dominance of these periodicals rather than the productivity of individual authors, as was pointed out earlier in this chapter.

Although the *M* corpus consists only of articles produced in connection with the research program of AFOSR, one might argue that its research front in fact represents the areas of concern to the research fronts of the entire field.

3.1.4 Epidemic Analysis

Epidemic analysis is of value in charting the rate of growth or decline in the population of researchers in a given field. In the present study the technique is applied to the entire field of information science insofar as it is represented by the main corpus and, individually, to the more productive specialties within that field. The population of authors for each field and subfield, and for each year during the period studied, is shown in Table 3.7. The procedure for charting the epidemic curve is as follows:

1. For each author, indicate in which years he published, the year in which he initially became infective, and the year of removal. An author is considered to have become infective in the year of his first publication listed in the AFOSR bibliography and to have become a removal in the year following his last publication in that list. In some cases, an author became reinfective subsequent to his removal; hence, the number of infections does not equal the total population.

2. Tally the number of new infectives I and removals R for each year.

3. Subtract the number of removals from the number of infectives, for each time period, to obtain the rate of change.

Table 3.7. *M* Corpus Author Population by Year

Subject	58	59	60	61	62	63	64	65	66	67
Abstracting and Indexing						2	4		3	
Aeronautics										3
Automation			1			7	1	9	19	15
Automation, General			1			2		5	4	3
Automation, Computers						2		2	10	10
Automation, Data Processing and Operations Research									1	1
Automation, Documentation						3	1	2	4	1
Banking and Finance									1	
Biological Sciences								1		
Bio. Sciences, Biology									2	3
Bio. Sciences, Biophysics								1		
Chemistry									1	
Chemistry, Analytic									1	
Education									1	1
Electricity and Magnetism				1		1	3	1	10	7
General Periodicals, U.S.					1					
Library Periodicals						3	1			1
Linguistics								1		
Literary and Political			1							1
Management					1					1
Mathematics						1			7	7
Medical Sciences, General			2	4	2		1	2	3	3
Philosophy	1						1	1	4	
Physics, General				1				1	3	2
Political Science										1
Psychology					1			2	1	2
Science					2	3	3		5	3
Sociology					1					
Sound Recording										1

4. Chart the rate of change over time.

5. Apply a similar procedure to the literature of each subcategory to obtain its epidemic curve.

The epidemic curve of corpus *M*, graphed by two-year periods, is shown in Figure 3.10. There is steady growth between 1959 and 1963, followed by a sharp decline in 1963 to 1965, after which the field entered an epidemic state for the last period studied. Epidemic curves for the most productive subcategories are shown in Figures 3.11 to 3.19. Among the individual subfields, the earliest activity is noted in the period 1959 to 1961 in three categories, namely, Electricity and Magnetism (Figure 3.15), Medical Sciences, General (Figure 3.17), and Science (Figure 3.19). In all, nine fields and subfields had sufficient activity to make possible a significant analysis on an individual basis. These were

Automation	Figure 3.11
Automation, General	Figure 3.12
Automation, Computers	Figure 3.13
Automation, Documentation	Figure 3.14
Electricity and Magnetism	Figure 3.15
Mathematics	Figure 3.16
Medical Sciences, General	Figure 3.17
Philosophy	Figure 3.18
Science	Figure 3.19

All nine fields show an initial rise followed by a sharp decline. Three fields never recovered from this decline, namely, Automation, General; Automation, Documentation; and Philosophy. The overall field of Automation did recover during the period 1965 to 1967, because the subfield, Computers, went into a highly epidemic state in that period.

The same pattern noted in Automation, that is, a decline during 1963 to 1965, followed by an epidemic during 1965 to 1967, was seen in the fields of Electricity and Magnetism, Mathematics, and Science.

Atypical of this pattern was the category of Medical Sciences, General, which went into its decline earlier than the rest but recovered before the overall epidemic began, and continued to rise to the end of the period studied.

There are thus three groups of categories, in terms of their epidemic activity.

Group 1 consists of those that had an early activity but died out before the beginning of the overall IS epidemic in 1965. This group includes Documentation and Philosophy (Figures 3.14 and 3.18).
Group 2 consists of those that declined during 1963 to 1965, but following that low point in overall IS activity, entered an epidemic state during 1965 to 1967. This group includes Computers (Figure 3.13), Electricity and Magnetism (Figure 3.15), Mathematics (Figure 3.16), and Science (Figure 3.19).
Group 3 consists of Medical Sciences, General (Figure 3.17), which went into its decline earlier than the overall decline, but rose at the time when others were declining, and then continued to climb during the general IS epidemic.

There are several interpretations possible for the pattern described by these three groups of subject categories. The shift

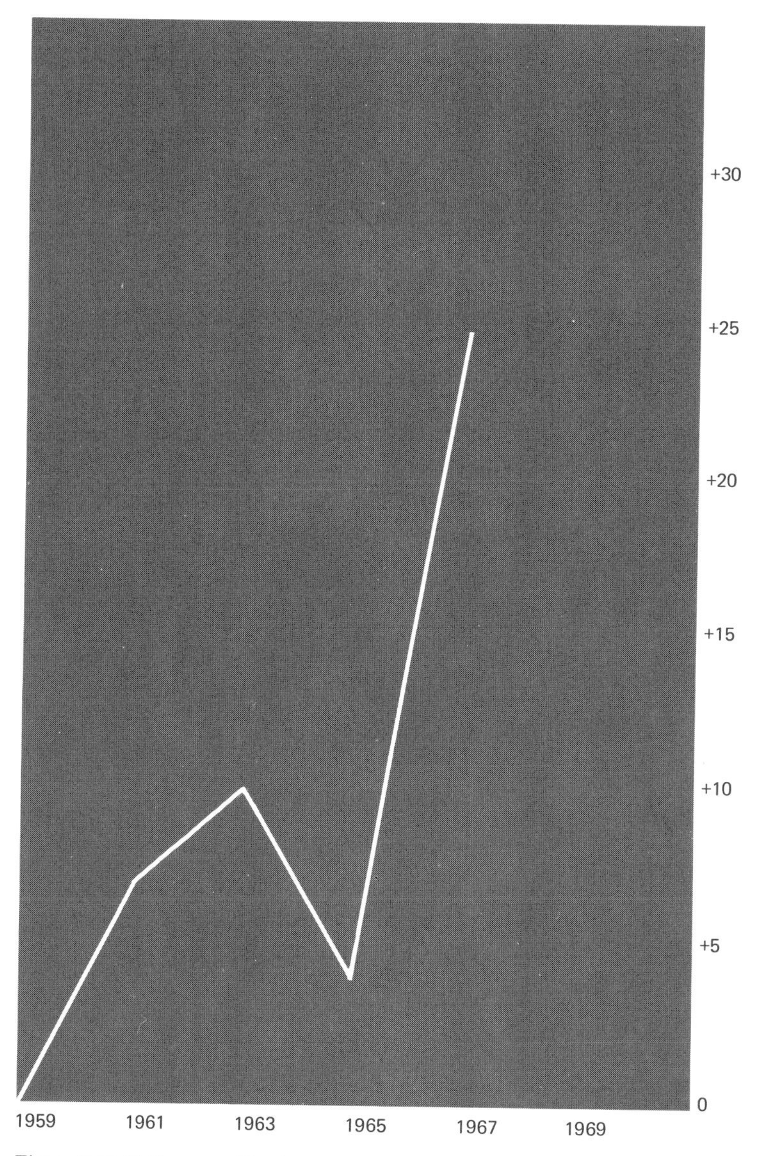

Figure 3.10 *M* Corpus epidemic curve

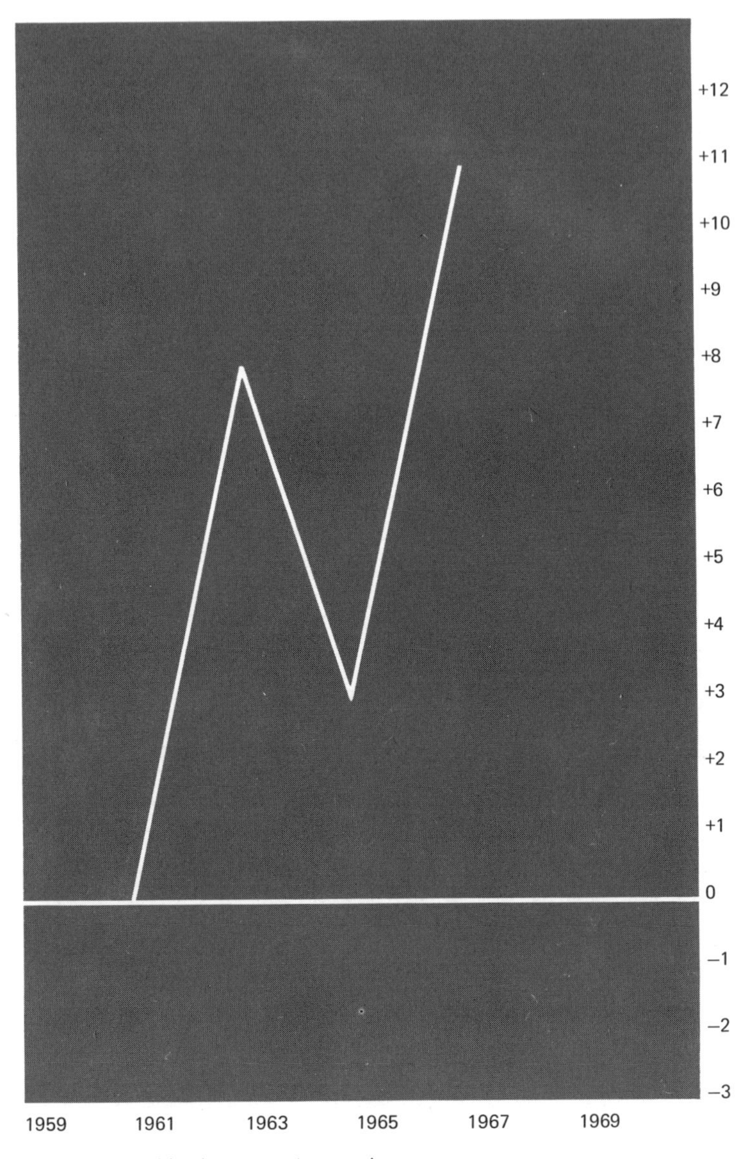

Figure 3.11 Epidemic curve—Automation

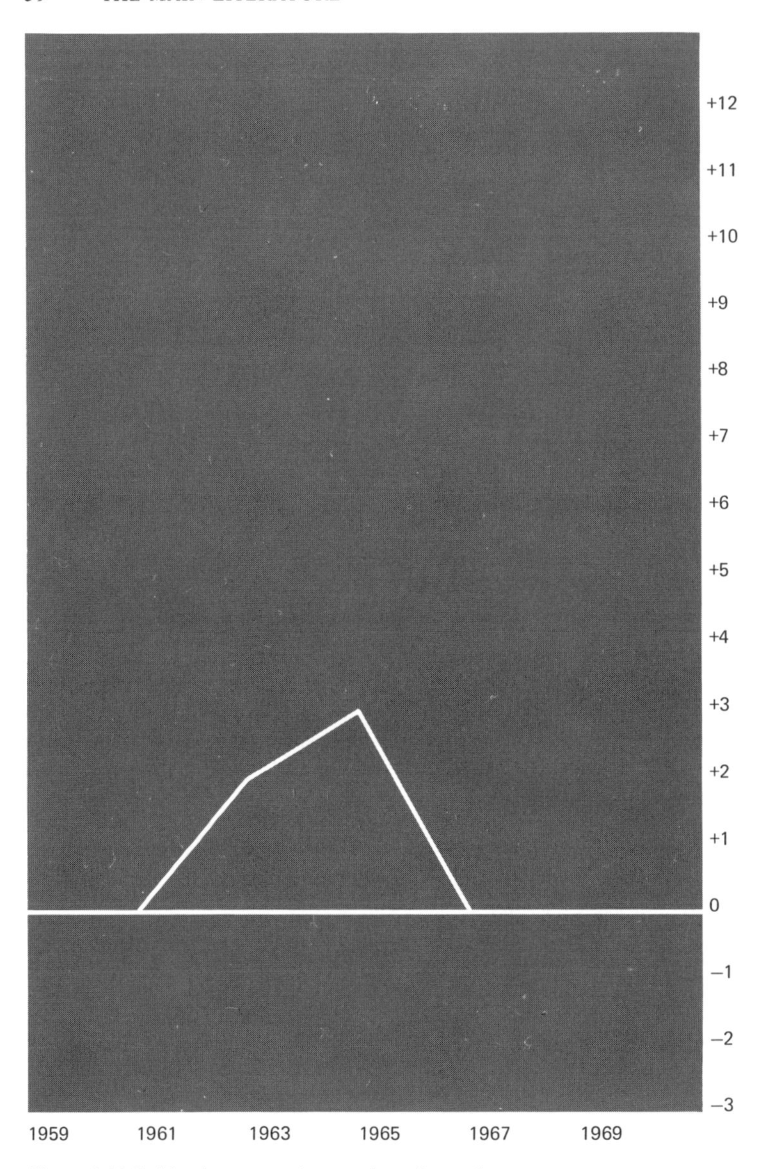

Figure 3.12 Epidemic curve—Automation, General

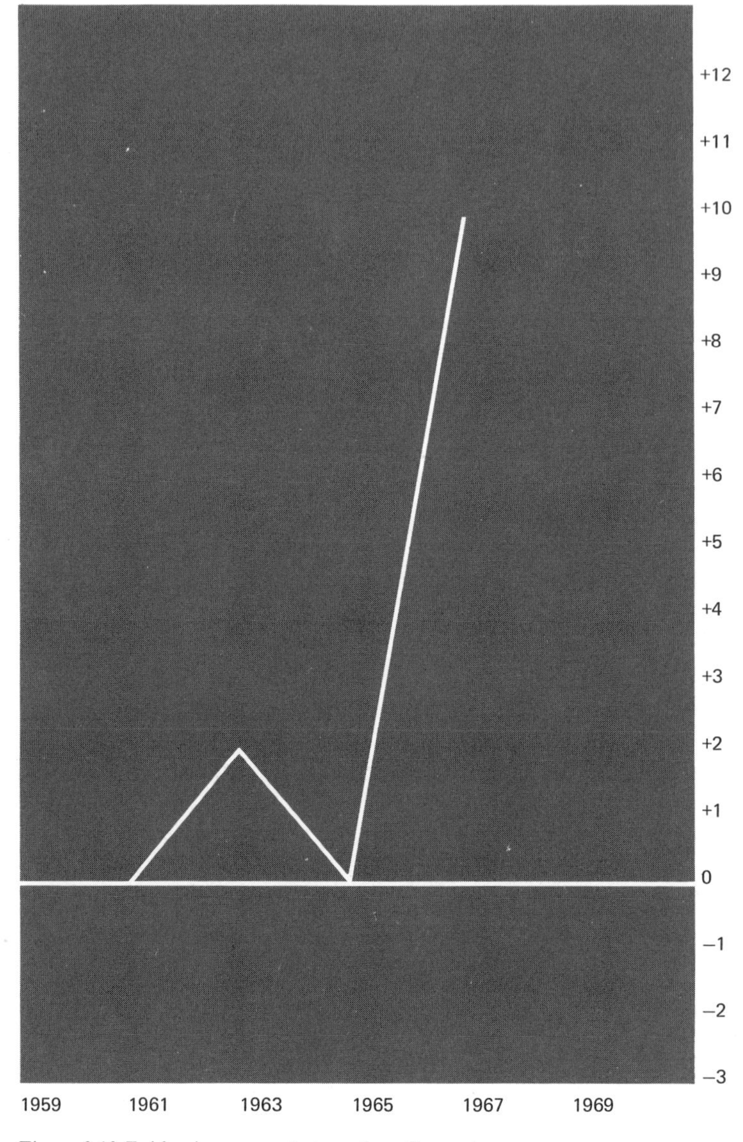

Figure 3.13 Epidemic curve—Automation, Computers

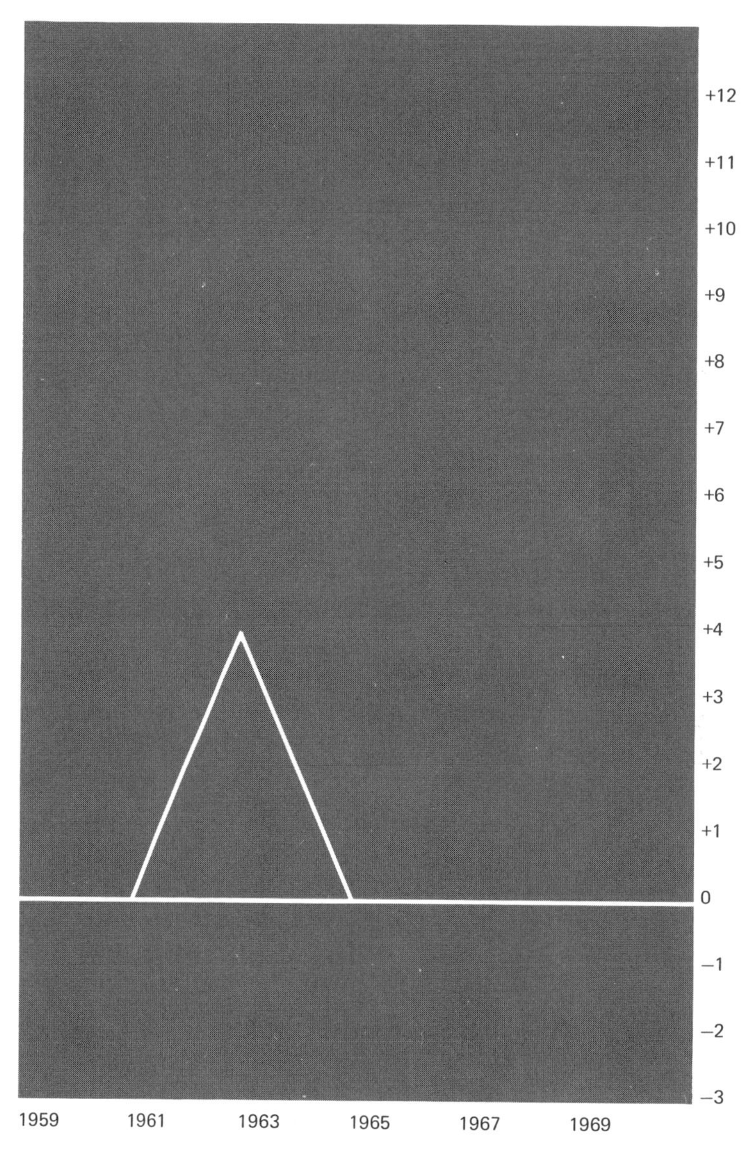

Figure 3.14 Epidemic curve—Automation, Documentation

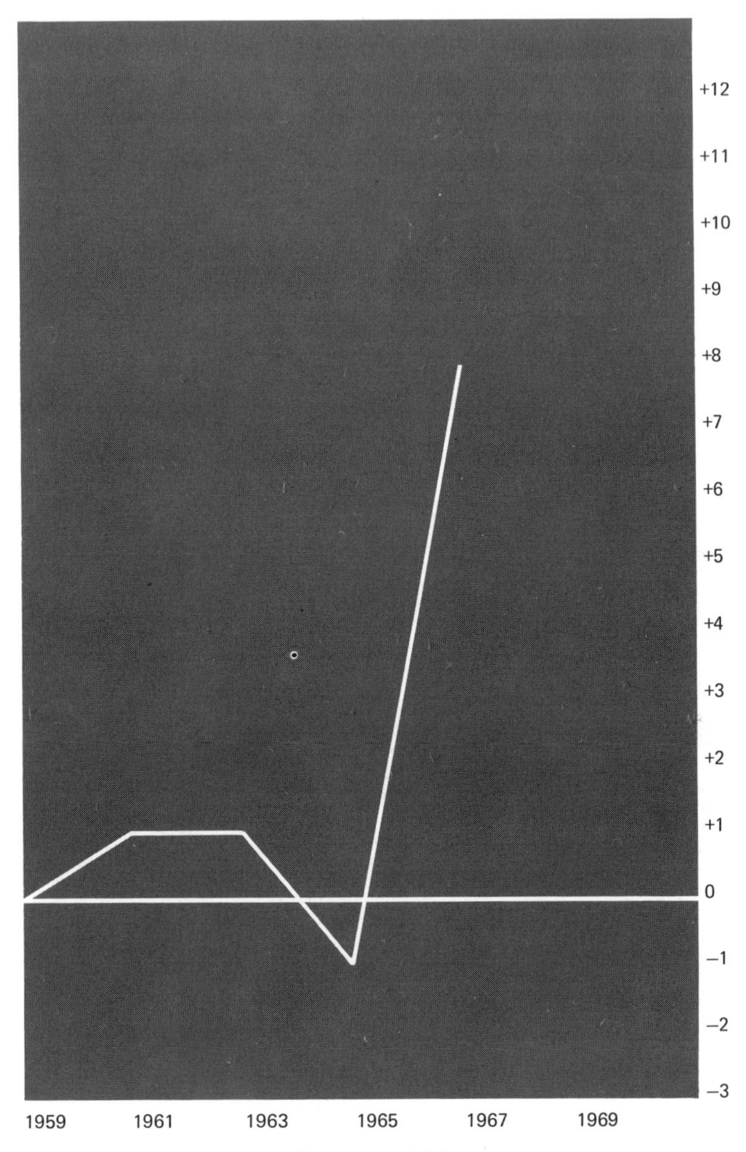

Figure 3.15 Epidemic curve—Electricity and Magnetism

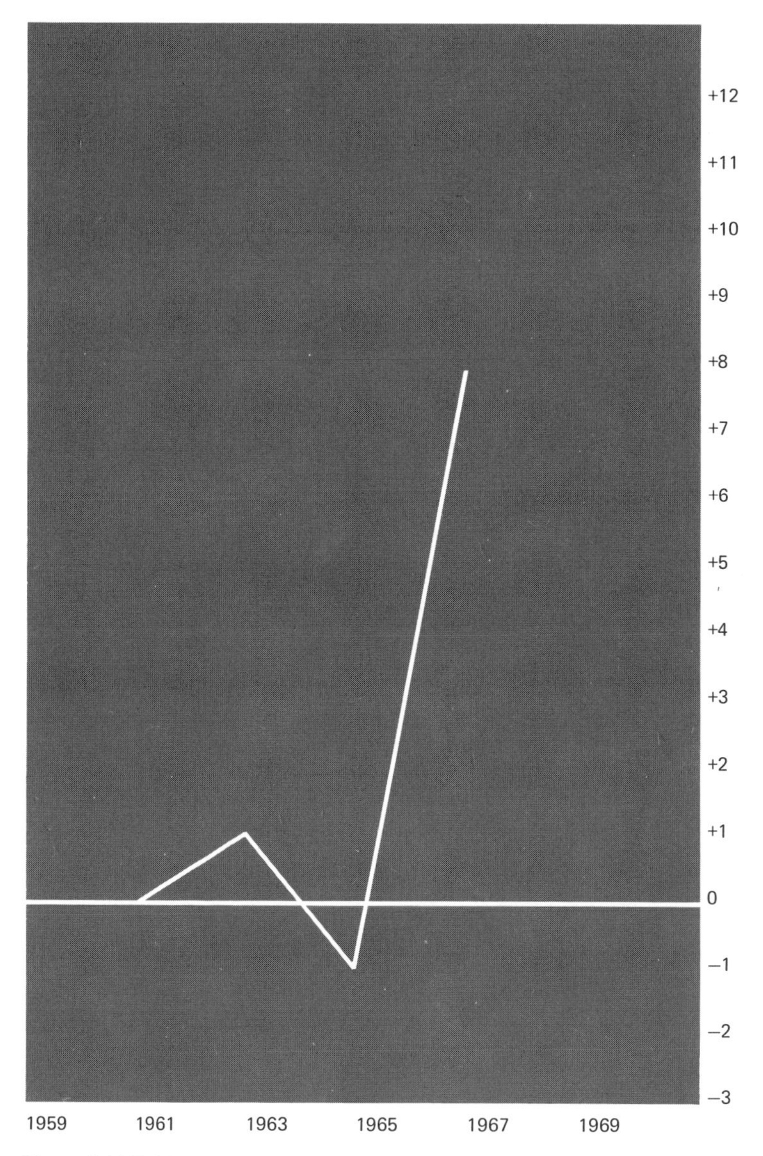

Figure 3.16 Epidemic curve—Mathematics

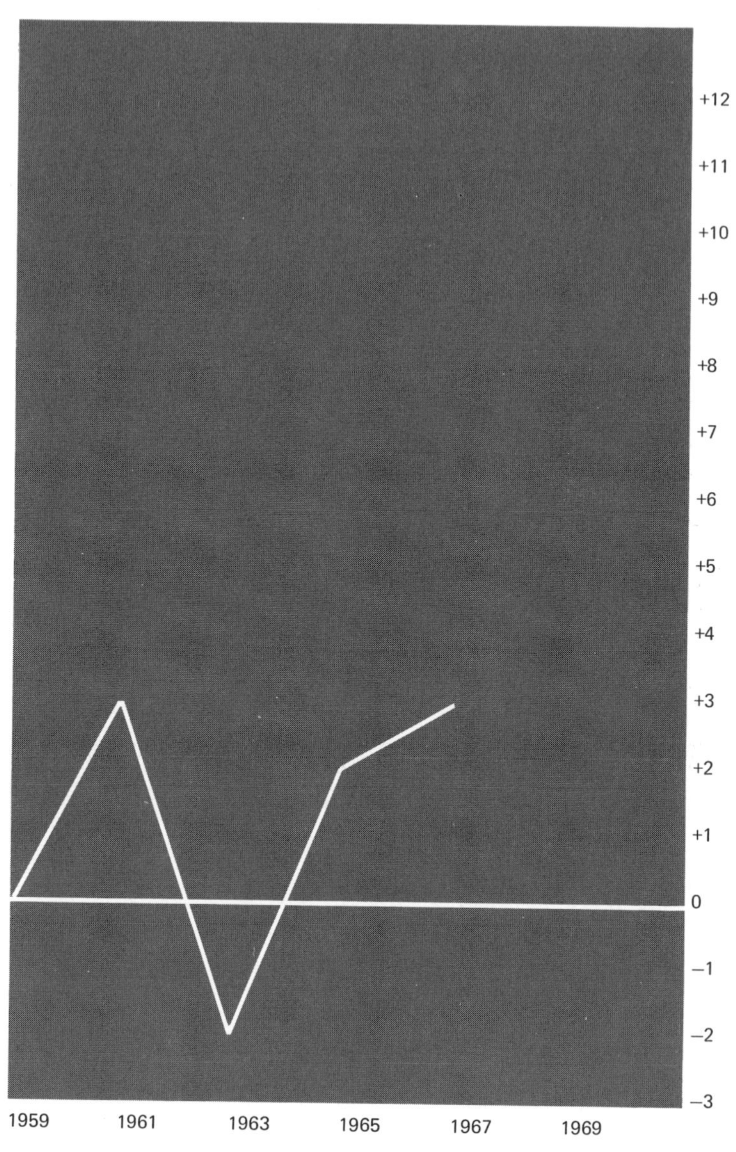

Figure 3.17 Epidemic curve—Medical Sciences, General

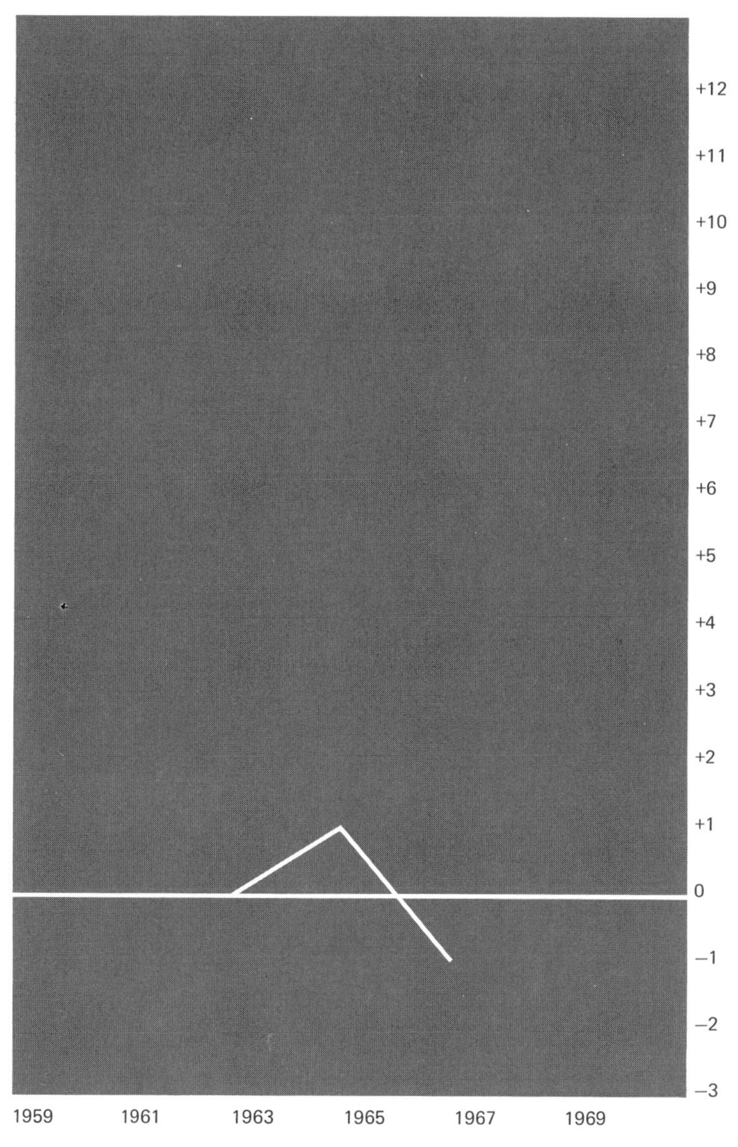

Figure 3.18 Epidemic curve—Philosophy

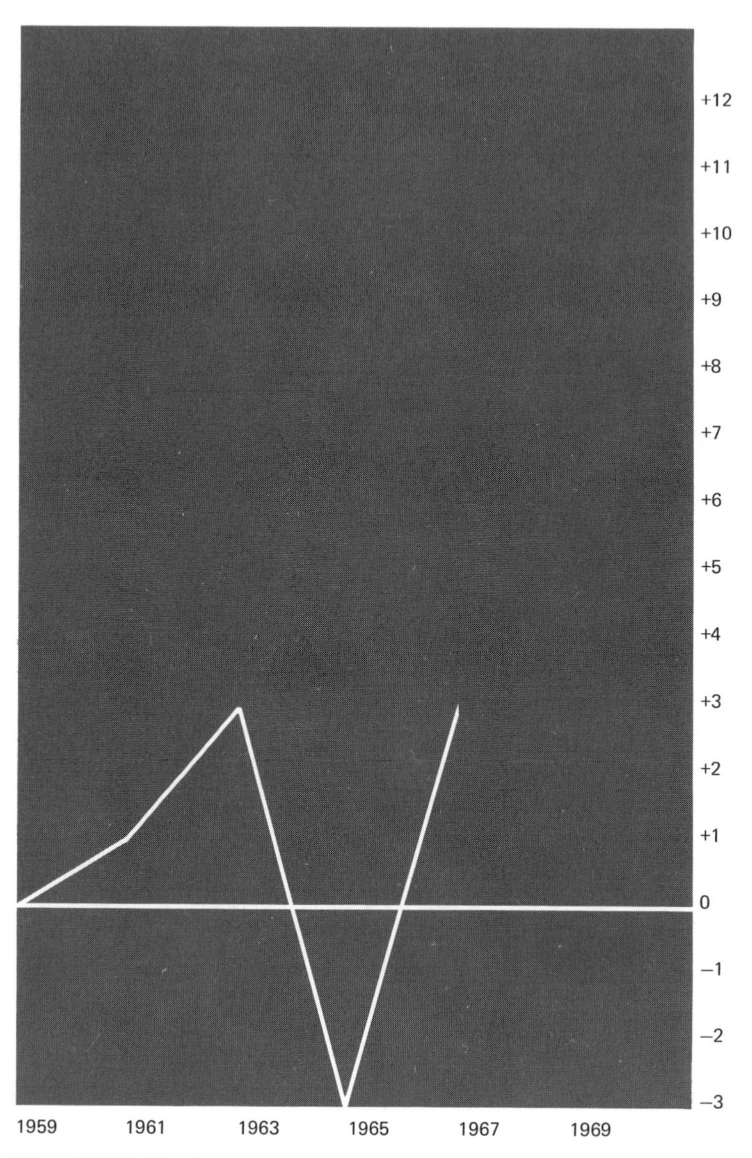

Figure 3.19 Epidemic curve—Science

in emphasis away from the fields of Documentation and Philosophy may well be related to the sponsor's decision to concentrate on theoretical research rather than on applications. But the continued growth of the computer and electrical engineering fields belies this explanation. It seems more likely that there is something that Documentation and Philosophy have in common that the other fields do not share. Both are in some sense external to the doing of science, even though philosophic speculation and analysis may precede and accompany scientific endeavor, and documentation both supports science and reports its results. The two activities, philosophy and documentation, may thus be centered, respectively, at opposite ends of the time continuum, before and after the actual scientific work.

Another explanation may be sought in the fact that a great deal of the interest in what we now call information science grew from the postwar interest in the creation of large-file systems and, also, in cybernetics. The former is a very practice-oriented problem, the latter a subject that seems to stimulate greatly such philosophic concerns as logic and the socioethical implications of science. It would appear likely that, after the initial explorations in both Documentation and Philosophy, researchers met with problems that could not be solved. Attention then apparently shifted to problems that could be posed in more formal terms and to the scientific disciplines that have the methods to solve them.

Finally, it is important to note that those fields that were in an epidemic state at the close of the period encompass most of the areas belonging to the research front, including Mathematics, Science, Engineering, and Computing. In spite of the fact that Documentation was not epidemic at the end of the period, its appearance in the research front suggests its continued importance.

3.2 The Support Literature

The value for a library in studying its own literature is apparent. But in order to understand the forces that shape that literature, it is necessary to go back to the subject fields from which its concepts are drawn and to derive some sense of the dynamics through which the transfer of information takes place. This may be done through a study of the sources *cited by* the literature of main concern.

Such a study has been performed here on the literature cited by the main corpus. The writings thus identified constitute a body of *support* literature, denoted here as the *S corpus*. In preparing it for analysis, the same criteria were used as in the preparation of the main corpus in that reports and other nonrefereed publications were excluded. But journal articles consisting of reviews were included, in an effort to assure that the *S* corpus would contain all contributions to IS that were recognized by authors of the main literature.

The *S* corpus contains 515 articles by 447 authors writing in 155 journals. These articles were cited by *M* corpus authors 627 times in all. Authors in *S* were cited an average of 1.4 times each.

The ratio of citations to authors is close to the ratio that Price found for citations to journals in the scientific literature generally, as mentioned in Chapter 2. The year-by-year production of the *S* corpus articles is shown in Figure 3.20,

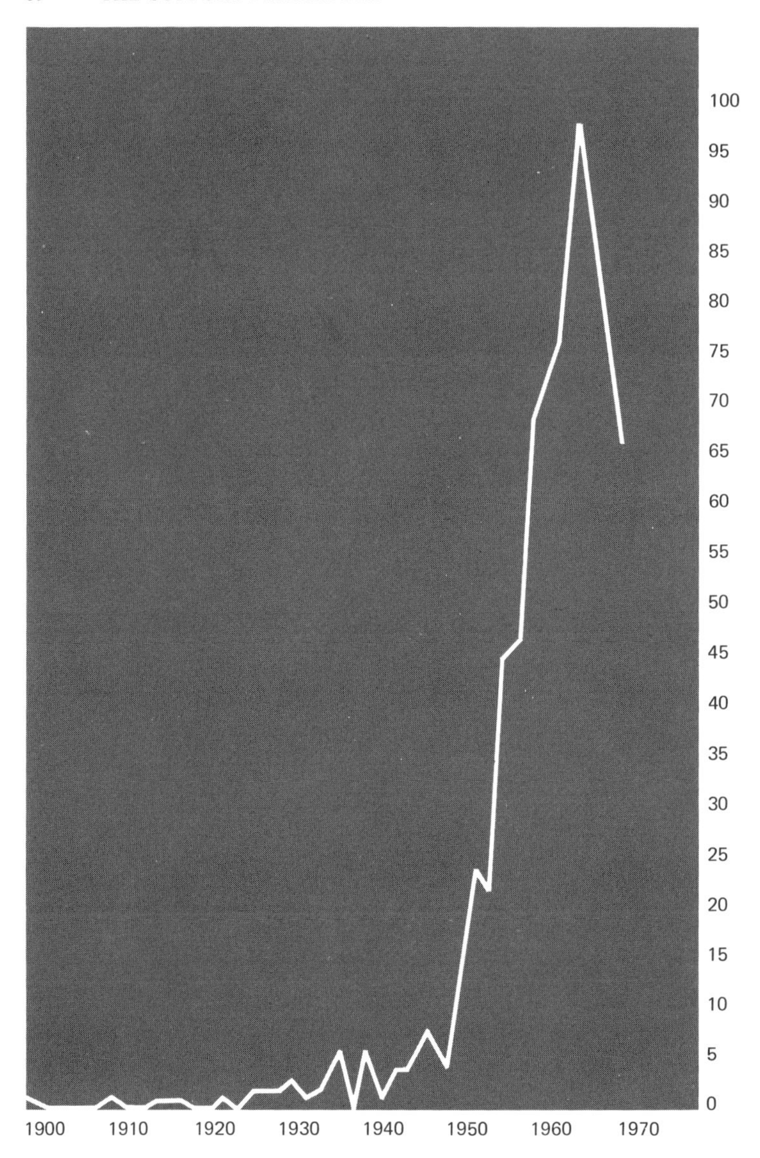

Figure 3.20 *S* Corpus articles

and that of highly productive categories in Figures 3.21 to 3.29. The categories of periodicals analyzed individually were

Automation	Figure 3.21
Automation, Computers	Figure 3.22
Biological Sciences	Figure 3.23
Electricity and Magnetism	Figure 3.24
Mathematics	Figure 3.25
Medical Sciences, General	Figure 3.26
Philosophy	Figure 3.27
Physics, General	Figure 3.28
Science	Figure 3.29

Price's comment, that the principal use of a journal article as measured by citations to it occurs in its first 10 years may be recalled in noting that the mass of citations from the IS literature were to publications in the previous decade. Notwithstanding this general tendency, there is in certain fields a sizeable number of citations to much earlier works. These include Biological Sciences (Figure 3.23), Mathematics (Figure 3.25), and Physics (Figure 3.28). Mathematics literature deserves special attention. As Figure 3.25 shows, there were certain earlier years, during the period 1936 to 1950, whose mathematics literature was cited with some degree of frequency. These years were separated by periods whose literature was not cited at all.

3.2.1 *S* Corpus Journals

The journals in the supporting corpus ranged in subject field from physical science (Acoustics) through social science (History) to the humanities (Philosophy) (See Table 3.8). An anomaly was noted in dealing with the great number of separate journals in this corpus, namely, that some organizations produce

Table 3.8. *S* Corpus Articles by Subject Field

Subject Field	Total	Subject Field	Total
Abstracting and Indexing Services	22	Mechanical Engineering	2
Anthropology	8	Chemical Engineering	1
Archeology	12	General Periodicals, U.S.	1
Architecture	1	Library Periodicals	24
Automation, General	12	Linguistics and Philology	11
Automation, Computer	29	Literary and Political Reviews	1
Automation, Documentation	3	Mathematics	117
Biology	5	Medical Sciences, General	50
Biophysics	5	Metallurgy	2
Physiology	17	Pharmacy and Pharmacology	2
Zoology	1	Philosophy	25
Chemistry, General	1	Physics, General	53
Chemistry, Biological	2	Political Science	1
Crystallography	1	Psychology	15
Electrochemistry	2	Public Administration	1
Education	5	Science	28
Electricity and Magnetism	42	Sound Recording and Reproduction	1
Engineering, General	1	Statistics	2
Instruments	2	Telephone and Telegraph	7

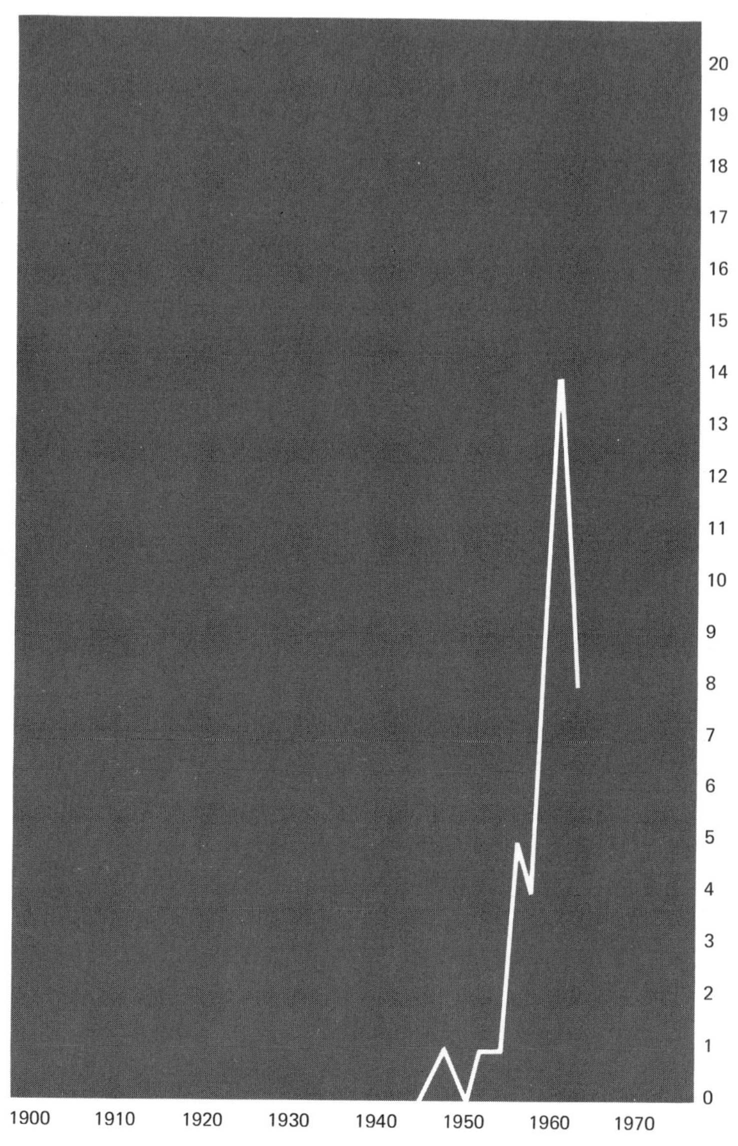

Figure 3.21 *S* Corpus—Automation

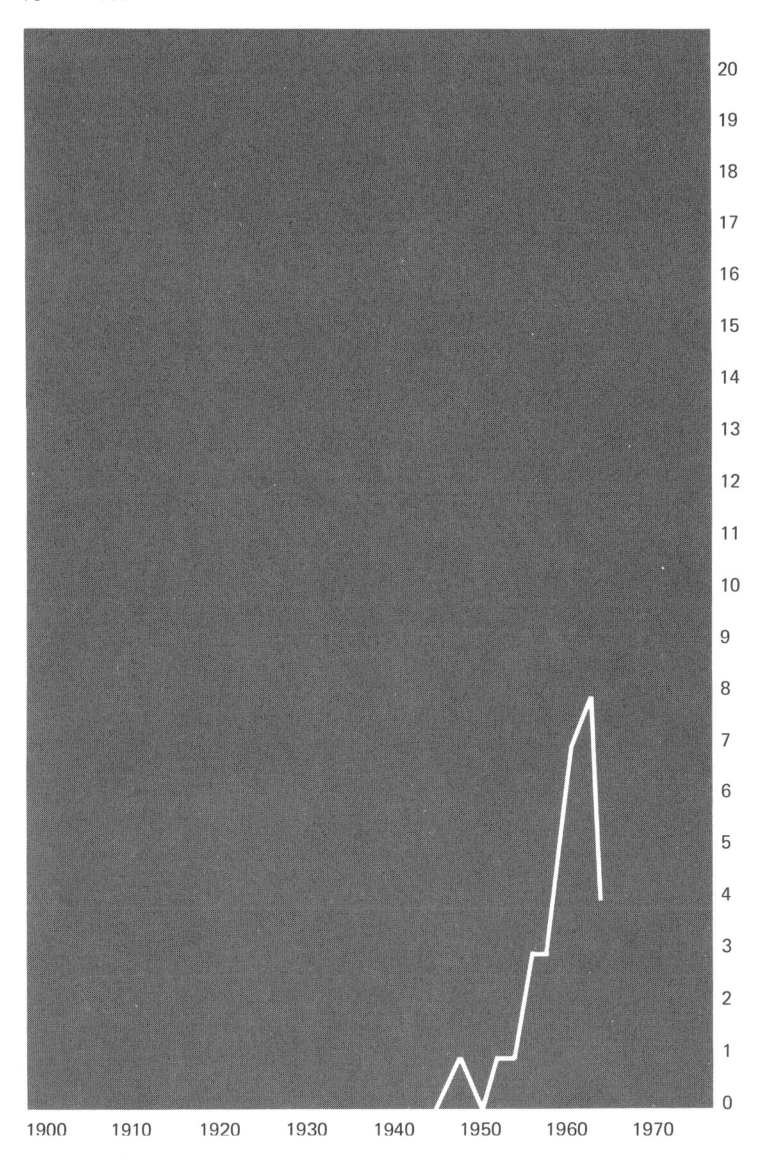

Figure 3.22 *S* Corpus—Automation, Computers

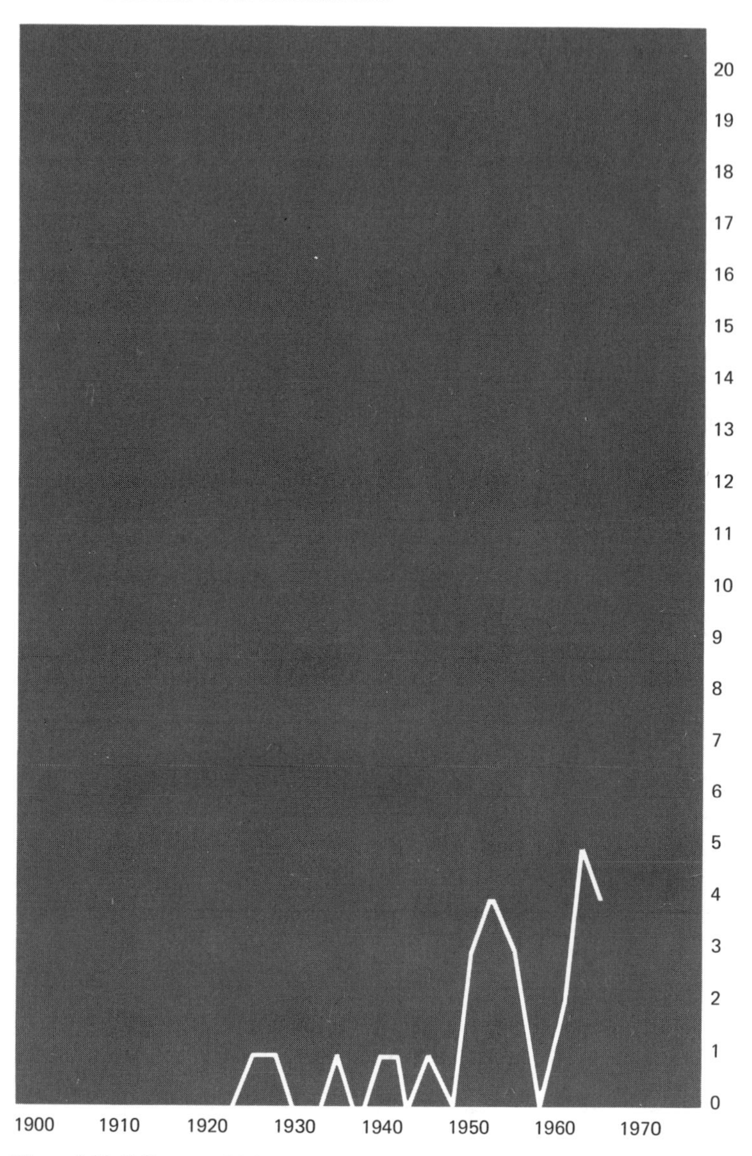

Figure 3.23 *S* Corpus—Biological Sciences

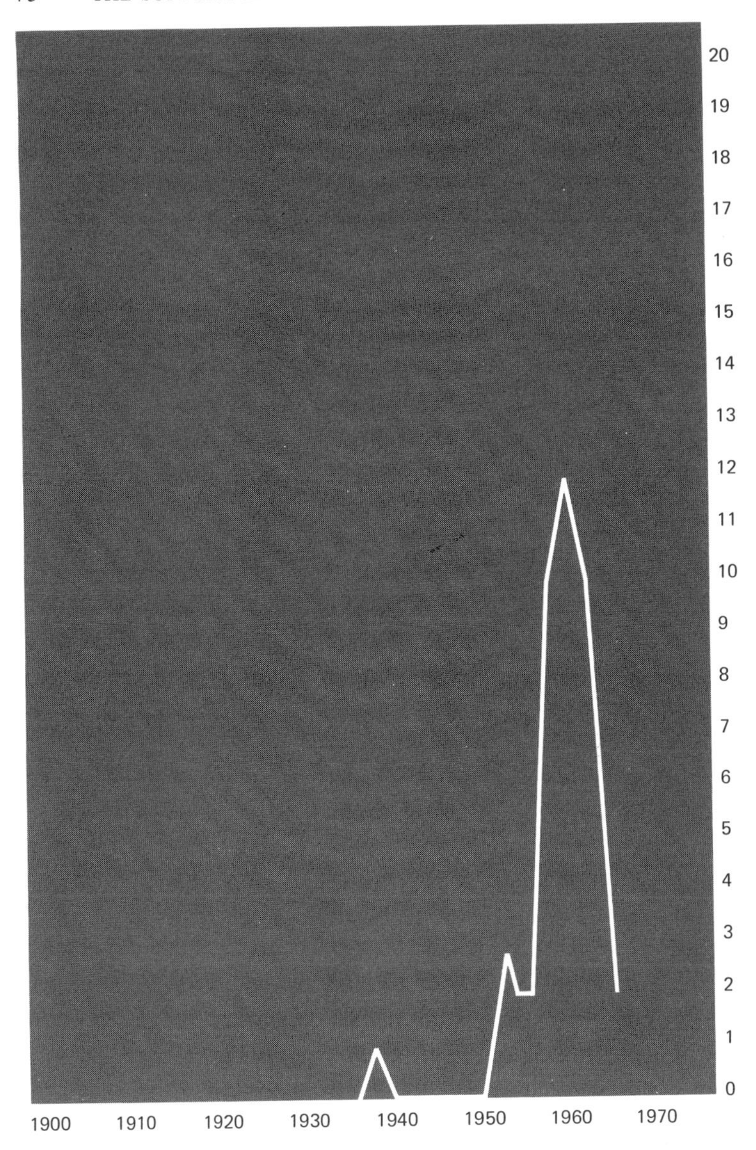

Figure 3.24 *S* Corpus—Electricity and Magnetism

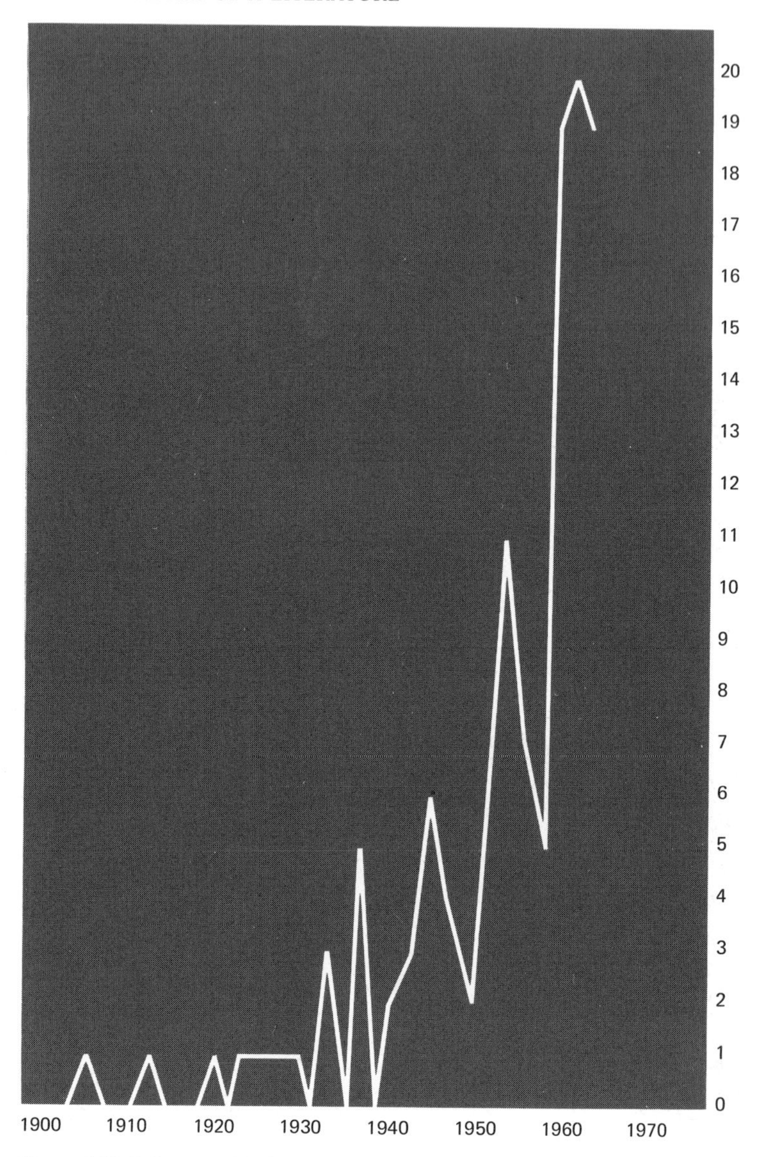

Figure 3.25 *S* Corpus—Mathematics

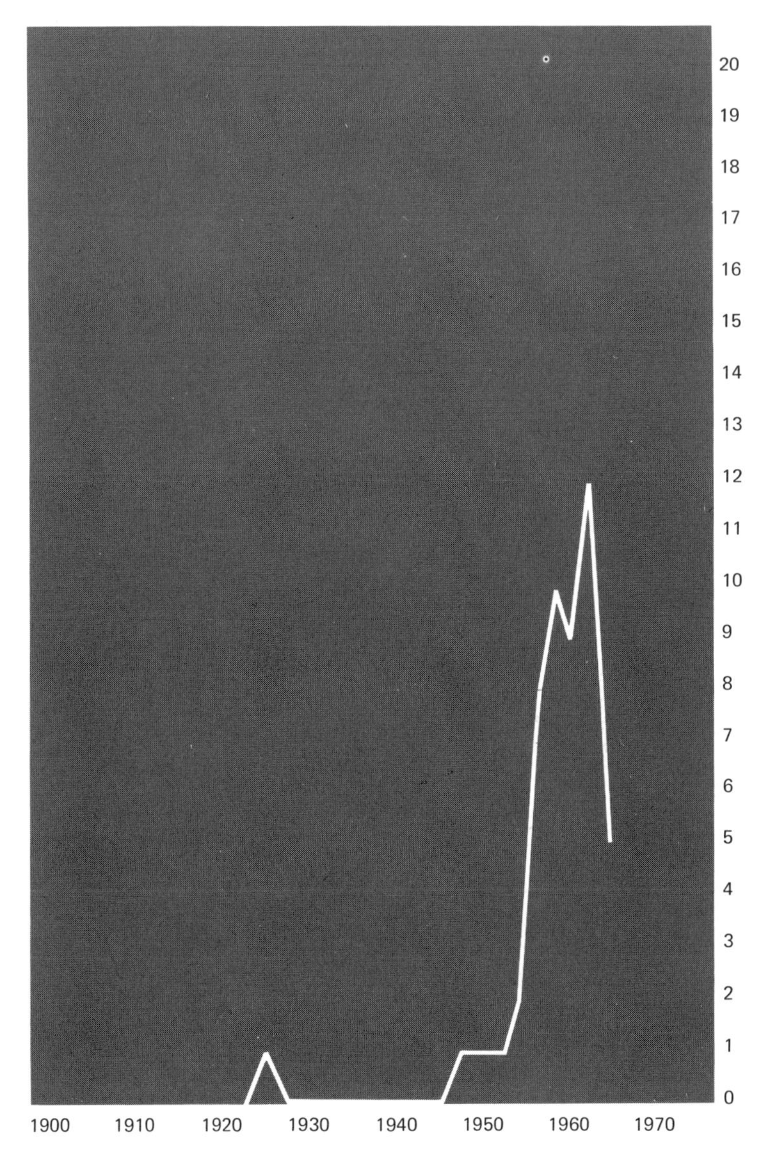

Figure 3.26 *S* Corpus—Medical Sciences, General

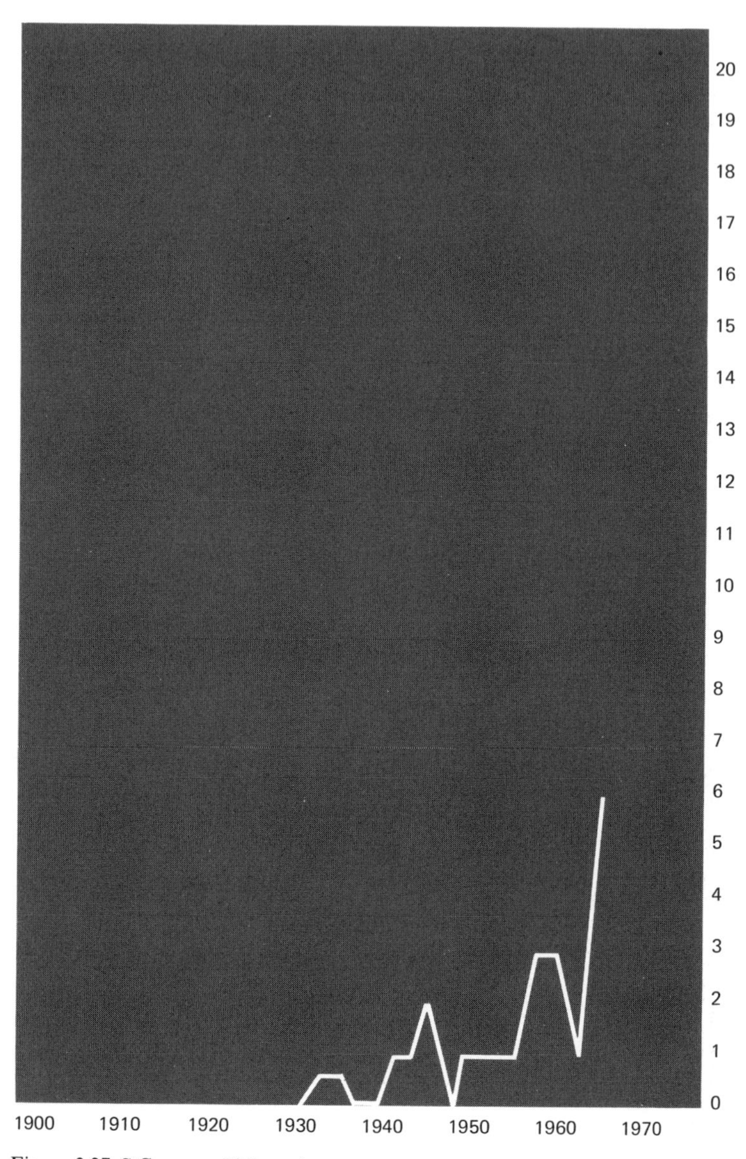

Figure 3.27 *S* Corpus—Philosophy

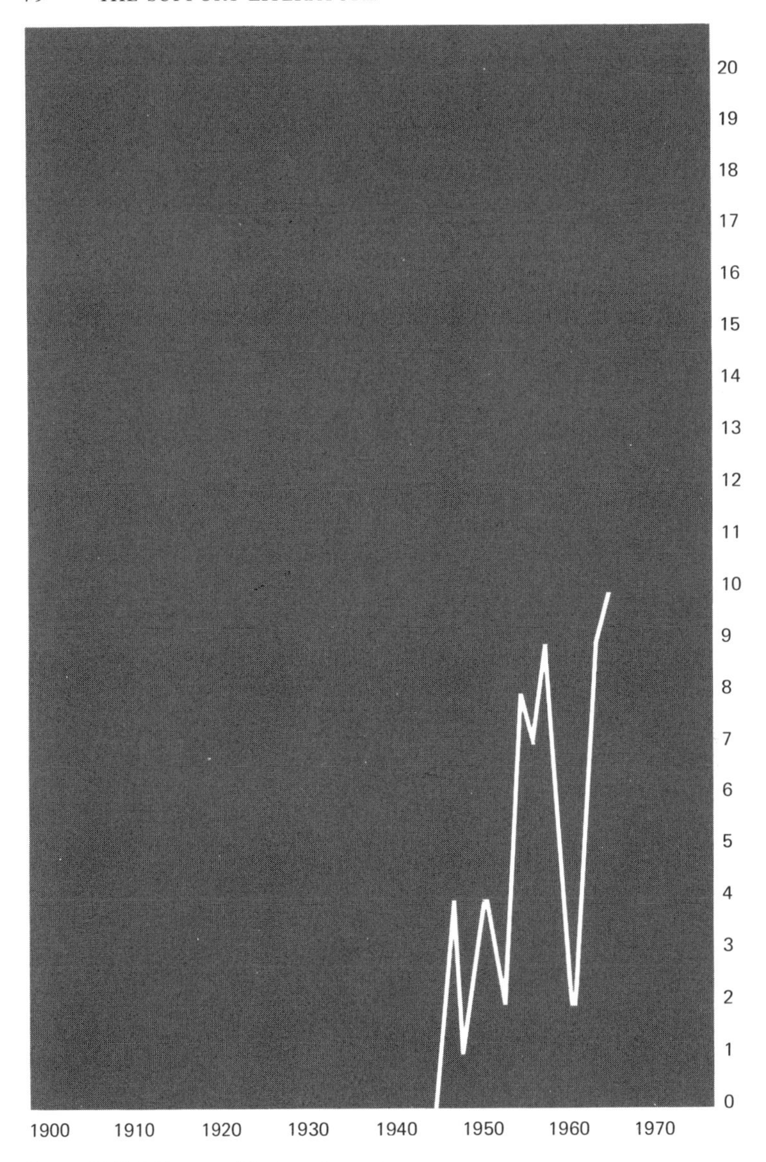

Figure 3.28 *S* Corpus—Physics, General

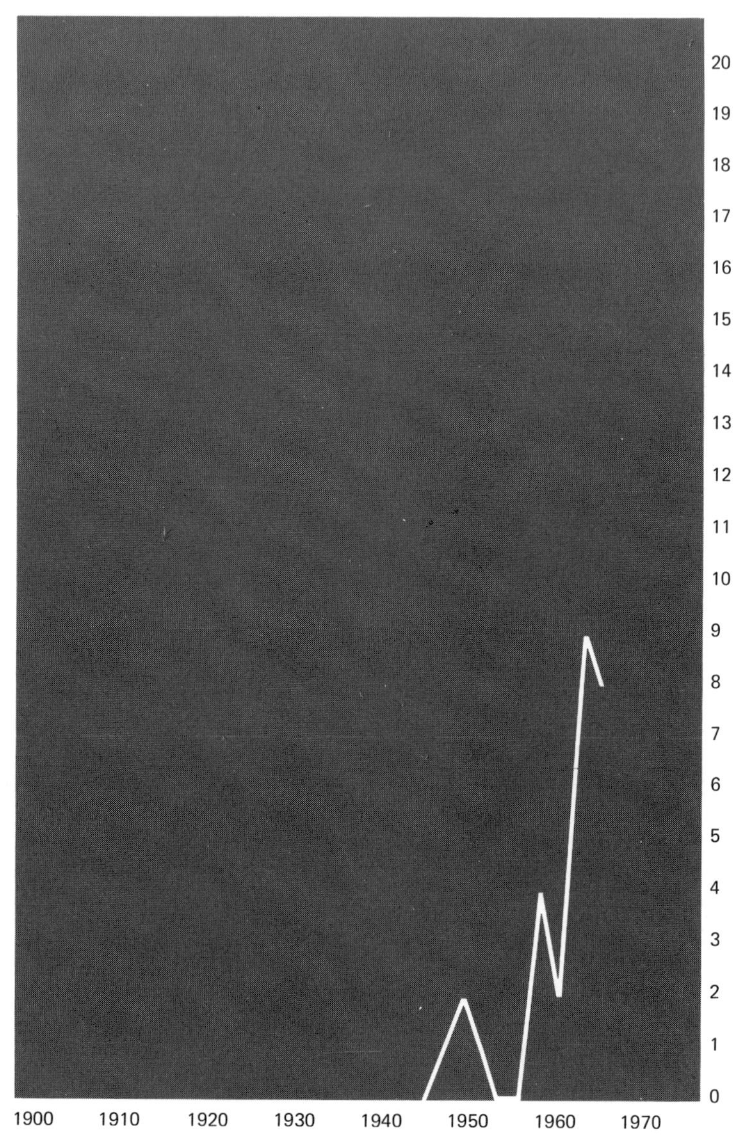

Figure 3.29 *S* Corpus—Science

a disproportionate number of journals, many or all of which might be considered in some sense as not really separate journals but as separate sections of a single journal. The result is a masking of the evidence of a Bradfordian distribution. If these journals (for example, the many *Transactions* of the IEEE) are grouped, the Bradford distribution is readily apparent in the dispersion of citations over journals in *S*. The distribution of citations by journals is seen in Table 3.9, with the Bradford partition shown in Table 3.10. As indicated there, the articles in the support literature were dispersed through the journals in

Table 3.9. Citations to *S* Corpus—Distribution

Journals	Citation	Total
1	52	52
1	34	34
1	21	21
1	19	19
1	15	15
2	14	28
1	12	12
1	11	11
1	10	10
1	9	9
3	8	24
3	7	21
4	6	24
5	5	25
12	4	48
7	3	21
17	2	34
14	2	28
79	1	79
155		515

a Bradfordian manner. The minimal nucleus consisted of one journal, cited 52 times. We find $T = 12$ for the distribution; therefore, all journals cited more than 12 times would be considered to be of a high order of "citedness." These include

Acoustical Society of America Journal
IEEE (all journals)
Annals of Mathematical Statistics
American Mathematical Society Journals
Journal of Speech and Hearing Research
Journal of Mathematics and Mechanics
American Documentation

These journals may be added to the core list of journals in the main corpus to form the set of the most useful literature sources for information science. The list now includes

Acoustical Society of America Journal
ACM Communications
American Mathematical Society Journals
Annals of Mathematical Statistics
IEEE (all journals)
Information and Control
Information Storage and Retrieval
Journal of Mathematics and Mechanics
Journal of Speech and Hearing Research

3.2.2 *S* Corpus Author Citedness

Citedness of *S* corpus authors is shown in Table 3.11. It should be remembered that this corpus contains only the items that were cited by the main literature and does not necessarily reflect the total citedness of these authors' works by the IS field at large. Therefore, it is not surprising that this distribution is non–Bradfordian. It is included here as an interesting example of nonapplicability of the law.

Table 3.10. Bradford Distribution of Citations to Journals in *S*

Zone	*J*	*C*	Ratio
1	1	52	
2	2	55	2.
3	3	48	1.5
4	4	47	1.3
5	7	54	1.8
6	9	49	1.3
7	13	51	1.4
8	23	52	1.8
9	38	52	1.7
10	55	55	1.4
155			
$b_m = 1.57$			

Table 3.11. *S* Corpus Author Citedness

Authors	Times Each Cited	Total
1	13	13
1	10	10
1	9	9
4	8	32
3	6	18
2	5	10
11	4	44
6	3	18
55	2	110
363	1	363
447		627

3.3 The Citing Literature

In Section 3.2 the study of the *S* corpus provided additional data about the processes occurring in the literature of information science and about the fields from which the main literature draws, the time span of publication of the support literature, and the journals of most value to IS research. But in order to provide adequate library service to a discipline, we should have some understanding of the direction of its research

Table 3.12. *C* Corpus Articles

Subject Field of	Year of Citing Journal				
Citing Journal	1964	1965	1966	1967	1968
Abstracting and Indexing	2	1		3	1
Automation	1	9		2	14
Automation, General		3		1	
Automation, Computers		2		1	14
Automation, Documentation	1	4			
Biological Sciences			1	4	2
Biological Sciences, Biology				4	1
Biological Sciences, Physiology			1		1
Education					1
Electricty and Magnetism	3	7	10	3	5
Engineering, Mechanical		1			
Library Periodicals	1	3		8	2
Management, Industrial		1			
Mathematics				6	1
Medical Sciences, General		3	6	2	8
Philosophy			2		
Physics, General			3	5	3
Political Science					1
Psychology	1			2	2
Science	1	2	2	3	
Sociology					2
Technology					1

in the future. Ideally, we ought to be able to predict the fields that will draw upon IS, the journals that will reflect the movement of IS concepts into those fields, and the time span during which this transfer will occur. The citing literature, denoted here as corpus *C*, was compiled by checking the *Science Citation Index* for references to the articles in *M*. The period covered in the indexes checked was from 1964 to the third quarter of 1968. The citation by *C* corpus articles of articles in *M* is shown in Table 3.12 by subject field and by year. The growth in the number of citations made each year by articles in *C* is shown in Figure 3.30.

3.3.1 *C* Corpus Journals

The distribution of citations from journals in *C* is shown in Table 3.13, and a Bradford partition of those journals in Table

Table 3.13. Distribution of Citations from *C* Corpus Journals

Journals	Number of Citations Made	Total
1	13	13
1	11	11
1	10	10
1	7	7
1	6	6
1	5	15
2	4	8
3	4	12
3	3	9
10	2	20
29	1	29
55		140

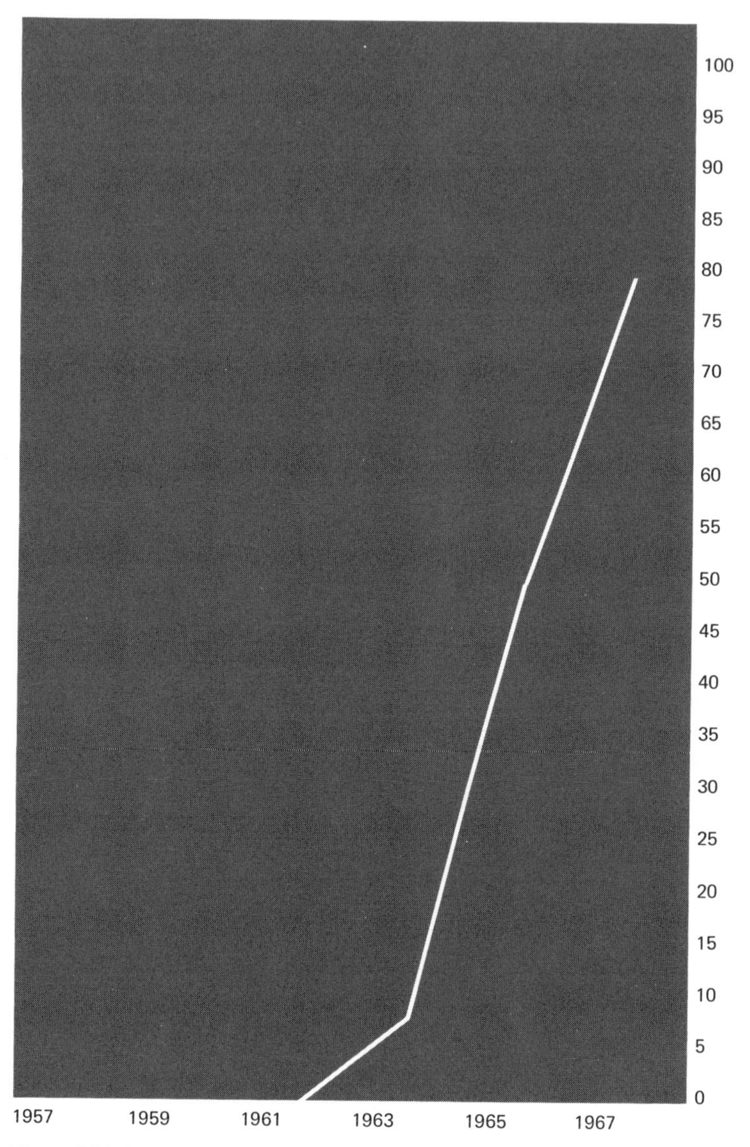

Figure 3.30 Citations to *M*, by year made.

Table 3.14. Bradford Distribution of Citations from *C* Corpus Journals

Zone	Journals	Number of Citations Made	Ratio
1	2	24	
2	3	23	1.5
3	5	23	1.7
4	7	23	1.4
5	15	24	2.1
6	23	23	1.5

3.14. The minimal nucleus of citing journals consists of one journal.
The partitioning formula yields $T = 7$, which identifies three journals as being of high productivity. These are

IEEE Transactions on Electronic Computers
Acoustical Society of America Journal
Numerische Mathematik

Adding these to our list of the most useful journals from *M* and *S*, we have now an optimal collection consisting of high-order journals of the three corpuses and denoted here as *N*. The collection consists of the following:

Acoustical Society of America Journal
ACM Communications
American Documentation
American Mathematical Society Journals
Annals of Mathematical Statistics
IEEE (all journals)
Information and Control
Information Storage and Retrieval
Journal of Mathematics and Mechanics

Journal of Speech and Hearing Research
Numerische Mathematik

These journals would be included in a collection that would reflect the flow of concepts through the literatures of supporting disciplines to the field of information science and, thence, to other disciplines that draw on IS. It is interesting to note that although several new periodicals were added as a result of analyzing the *S* and *C* literatures, these journals represent subject areas belonging to the research front of *M*. From this we may infer that the field of information science, albeit made up of several subject disciplines, seems to be self-contained in that it draws from and supports those disciplines that make up its own major areas of activity.

4
Summary and Conclusions

As stated in Chapter 1, this work is based on the assumption of certain regularities in patterns of authorship, publication, and citation of literatures, the analysis of which can yield principles for the effective management of libraries. Chapter 2 introduced three techniques that previously have been used for the analysis of the *static* properties of literatures; namely, Bradford analysis, citation tracing, and bibliographic coupling. Epidemic theory was introduced, on the other hand, as a means for the analysis of literature *dynamics*. These four techniques were integrated into a general method that was applied to a corpus of information science literature, augmented by the literature it cites and by publications citing it.

In the analysis of the statics of the main literature, it was possible to identify a minimal nucleus of *authors* on the basis of their productivity. A minimal nucleus of *journals* in the main literature was also identified. Because there were no journals of high productivity as defined by the Zipf-Booth-Goffman formula, the five journals in the nucleus would be selected for a library serving the reseachers in information science.

Using the techniques of Price, a research front for IS was identified, consisting of 32 articles, which represent the "growing tip" of research activity. Application of methods of Goffman and Kessler yielded a classification of the articles into ten distinct classes, implying the existence of ten small disjunct research fronts, rather than of a single front.

In analyzing the dynamics of the main literature, the use of epidemic theory showed that there were nine fields and subfields that at some time during the period studied had experienced significant growth, six of which might be said to have been in an epidemic state at the close of the period. The six, moreover, belonged to the research front, indicating that these fields had not reached a peak of activity and would continue to be fruitful areas of research for some time.

The support literature, that is, the publications cited by the main corpus articles, was also studied by the Bradford technique to identify the journals in the minimal nucleus and in the high productivity zone. This allowed us to add a number of journals not previously identified in the study of the main literature as being of high interest to IS research. Price's generalization concerning the shortness of a journal article's useful life was, in general, borne out. Similar analysis of the literature that cites the main corpus identified one additional journal for our list of high-value publications.

The set of journals identified from the three literatures was found to be representative of the subject areas in the research front of *M*, thus indicating that, although IS as defined here involves many disciplines, it appears to be self-contained, since it both draws upon and contributes to those disciplines in its own major areas of interest.

4.1 Use of the Method for Library Planning

The techniques applied, with the exception of Bradford's, were first developed in the context of the study of individual scientific literatures, rather than for use in the management of libraries. They have been applied here to the analysis of information science literature as a demonstration that they could indeed be useful in the processes of library acquisition and classification. The significance of the experiment lies not so much in the data derived about the discipline of IS, as in what it may contribute to the development of a general method of literature analysis for use in libraries. A great deal of work must be done to develop the procedure for further integration of these techniques and for the measurement of their validity.

Although the techniques are now being used primarily in the study of individual literatures for the purpose of gaining a scientific knowledge about the growth of science itself, they are also being applied experimentally in at least one major research library in this country, the Allen Memorial Medical Library, Cleveland, Ohio.

4.2 Procedure for Application of the Method to Libraries

The purpose of this section is to outline a procedure for application of the method to the problems of library management, especially as regards the management of subject literatures, and to suggest some alternative modes of approach within the general method.

The model used in the discussion is the research library, but this is not to suggest that the method's value is necessarily limited to that kind of application. The reasons for choosing the research library setting are that (1) it is clearly a type of library in which the nature of the collection is of paramount importance, and (2) the techniques employed here have been tested as they apply to scientific literature, but doubts remain about their applicability to the literatures of other disciplines.

The problems that have to do primarily with library

management of literatures are (1) identification and acquisition, (2) classification and storage, and (3) retrieval and dissemination. Of these, retrieval is treated here as a special case of acquisition; whereas the initial acquisition process consists of identifying items from the entire world's literature, retrieval on the other hand is the identification and acquisition of items from that set of documents that the library has already collected. No separate discussion of retrieval is included here.

On the basis of the experimental evidence derived from this study, a general method is suggested here for the application of these information-scientific techniques to library management. The method consists of seven steps in two stages. In the first stage, the static properties of the literature are established. The second stage develops its dynamic properties. The steps within these two stages are as follows:

Stage One

Step One.
Determine the frequency of the library's use by members of its clientele. This may be done in a number of ways, the simplest being on the basis of the kind of library-use statistics conventionally gathered. However, techniques should be devised to reflect the in-house use of materials in open-stack libraries.

Step Two.
Establish for the distribution a minimal nucleus and the transition point T between high- and low-frequency users. (The applicability of the Bradford technique to library circulation has been demonstrated by Goffman and Morris 1970. In such an application as is suggested here, however, a careful

distinction ought to be made between the kinds of use for which the library is established and incidental use which, although quite possibly legitimate, is not part of the library's purpose. The distinction may not be important in all cases. However, in many situations, such as in industrial research organizations, the library must be sensitive to the stated or implied purposes that management has in mind in providing support. Statistics of the library's use that are unrelated to the library's true purpose may result in an invalid assignment of people to the nuclear user group. Further correction might well be introduced to cover the cases where library users are, in fact, acting as agents for others. This might include the case of a librarian from another library borrowing for his clients, or a student borrowing for his teacher, or vice versa).

Step Three.
Determine the subject areas of interest to the high-frequency users. Any one of several techniques might be employed, for example, a user/interest profile such as is constructed for use in selective dissemination systems.

Step Four.
Analyze each subject area thus identified, as demonstrated in Section 3.3, using the research front concept and the techniques illustrated for its identification, then classifying the research front into component subject areas of activity.

Step Five.
On the basis of the foregoing, establish an optimal collection N of periodicals representing the research front of each field of interest of each high-frequency user.

Step Six.
Take the union of these individual optimal collections as the optimal periodicals collection for the entire library.

Stage Two

Step Seven.
Perform epidemic analyses of each population involved. These analyses can be used to record and to predict (1) changes in the constitution of the user group; (2) changes in the population of users of, and contributors to, each of the disciplines; (3) changes in the research front; (4) changes in the optimal periodicals collection; (5) changes in the support literature; and (6) changes in the citing literature.

Step Seven is necessary for keeping the collection current once it is established, by detecting changes in the values of the basic parameters.

This method can be used to establish the optimal collection in terms of both the static and the dynamic aspects of the literature. Thus, librarians can use it as a guide to planning, keeping in mind that the exact nature of its implementation will depend upon the library's purpose and functions as well as its budget. For example, some libraries may wish to establish acquisition policies in strict adherence to this method, while others may wish to acquire larger collections but to display the optimal collection in the most convenient location. Still others may wish to display only the minimal collection. The choice of periodicals from each of the three corpuses—main, support, and citing—may be weighted in recognition of the library's purposes. For example, a library whose principal purpose is to support the dissemination of ideas in the main corpus may weight more heavily the citing literature, which serves the cause of such promulgation, whereas a library whose sole purpose is to support the research efforts of the main corpus writers might well choose to weight more heavily the periodicals in the support corpus.

The building of library collections in a period of great changes, both in the scientific process and in the nature of its applications, requires more than a historical understanding of how ideas affect ideas. Today's specialized scientific collection, tailored to the current needs of a user population, must be altered on a continual basis to prepare for, rather than merely to reflect in a historical way, the changes in those needs. The method developed in this study seems to offer operational means of doing so.

Bibliography

AFOSR (Air Force Office of Scientific Research), Information Sciences Directorate. 1968. *Information Sciences—1967.* Arlington, Va.: Office of Aerospace Research.

Booth, Andrew. 1967. "A 'Law' of Occurrences for Words of Low Frequency." *Information and Control* 10: 386–393.

Borko, Harold. 1967. *Automated Language Processing*. New York: John Wiley & Sons.

Bradford, Samuel Clement. 1934. "Sources of Information on Specific Subjects." *Engineering* (London): January 26, 1934. 137: 85–86.

———. 1948. *Documentation*. London: Crosby Lockwood & Son.

Brookes, B.C. 1969. "Bradford's Law and the Bibliography of Science." *Nature* 224: 953–956.

Chicorel, Marietta, ed. 1967. *Ulrich's International Periodicals Directory*. New York: R.R. Bowker.

Fairthorne, Robert A. 1961. "Basic Postulates and Common Syntax." *Information Retrieval and Machine Translation* 2. Reprinted in Fairthorne, Robert A. *Towards Information Retrieval*. London: Butterworth & Co.

———. 1969. "Empirical Hyperbolic Distributions (Bradford–Zipf–Mandelbrot) for Bibliometric Description and Prediction." *Journal of Documentation* 25: 319–343.

Fenichel, Carol 1969. *Citation Patterns in Information Science*, Master's thesis, Drexel University, Philadelphia, Pennsylvania, June 1969.

Gardiner, G.L. 1969. "The Empirical Study of Reference." *College and Research Libraries* 30: 130–155.

Goffman, William. 1966. "Mathematical Approach to the Spread of Scientific Ideas—the History of Mast Cell Research." *Nature* 212: 449–452.

———. 1969. "An Indirect Method of Information Retrieval." *Information Storage and Retrieval* 4: 361–373.

———. 1970. "A General Theory of Communication." In *Introduction to Information Science*, Tefko Saracevic, ed. New York: R. R. Bowker.

———. *Mathematics of Information Science* . To be published.

Goffman, William, and Morris, Thomas G. 1970. "Bradford's Law and Library Acquisitions." *Nature* 226: 922–923.

Goffman, William, and Newill, Vaun A. 1964. "Generalization of Epidemic Theory: An Application to the Transmission of Ideas." *Nature* 204: 225–228.

———. 1967. "Communication and Epidemic Processes." *Proceedings of the Royal Society, A* 298: 316–334.

Goffman, William, and Warren, Kenneth S. 1969. "Dispersion of Papers Among Journals Based on a Mathematical Analysis of Two Diverse Literatures." *Nature* 221: 1205–1207.

Hulme, E.W. 1923. *Statistical Bibliography in Relation to the Growth of Modern Civilization.* London: Grafton and Company.

Kessler, M.M. 1963. "Bibliographic Coupling Between Scientific Papers." *American Documentation* 14: 10–25.

———. 1967. "The On-Line Technical Information System at M.I.T.—Project TIP." *IEEE International Convention Record, 1967*, Part 10: 40–43.

Kochen, Manfred. 1969. "Stability in the Growth of Knowledge." *American Documentation* 20: 186–197.

Leimkuhler, Ferdinand F. 1967. "The Bradford Distribution." *Journal of Documentation* 23: 197–207.

Lowes, John L. 1927. *The Road to Xanadu.* Boston: Houghton Mifflin Co.

Paisley, William J. 1968. "Information Needs and Uses." In *Annual Review of Information Science and Technology* , vol. 3, Carlos A. Cuadra, ed. Chicago: Encyclopaedia Britannica.

Pritchard, A. 1969a. *Statistical Bibliography: An Interim Bibliography.* London: North–Western Polytechnic, School of Librarianship.

———. 1969b. "Statistical Bibliography or Bibliometrics?" *Journal of Documentation* 25: 358–359.

Price, Derek J. de Solla. 1961. *Science Since Babylon.* New Haven: Yale University Press.

———. 1963. *Little Science, Big Science.* New York: Columbia University Press.

———. 1965. "Networks of Scientific Papers." *Science* 149: 510–515.

Rawski, Conrad H. 1971. "The Scientific Study of Subject Literatures." In *Proceedings of a Symposium in the Health Sciences: Working to the Future*, Robert G. Chesier, ed. Cleveland, Ohio: Cleveland Medical Library Association (11000 Euclid Avenue, Cleveland, Ohio 44106).

Science Citation Index. Philadelphia: Institute for Scientific Information, 325 Chestnut Street.

Selye, Hans. 1965. *The Mast Cells.* London: Butterworth & Co.

Stevens, Mary Elizabeth; Guiliano, Vincent E; and Heilprin, Laurence B; eds. 1965. *Statistical Association Methods for Mechanized Documentation.* Washington, D.C.: National Bureau of Standards.

Vickery, B.C. 1968. "Bradford's Law of Scattering." *Journal of Documentation* 4: 198–203.

Ziman, J.M. 1969. "Information, Communication, Knowledge." *Nature* 224: 318–324.

Zipf, G.K. 1935. *Psychobiology of Language.* Boston: Houghton Mifflin Co.

———. 1949. *Human Behaviour and the Principle of Least Effort.* Cambridge, Mass.: Addison–Wesley Publishing Co.

Index